Transonic Aerodynamics
Problems in Asymptotic Theory

Frontiers in Applied Mathematics

Frontiers in Applied Mathematics is a series that presents new mathematical or computational approaches to significant scientific problems. Beginning with Volume 4, the series reflects a change in both philosophy and format. Each volume focuses on a broad application of general interest to applied mathematicians as well as engineers and other scientists.

This unique series will advance the development of applied mathematics through the rapid publication of short, inexpensive books that lie on the cutting edge of research.

Frontiers in Applied Mathematics

Vol. 1	Ewing, Richard E., The Mathematics of Reservoir Simulation
Vol. 2	Buckmaster, John D., The Mathematics of Combustion
Vol. 3	McCormick, Stephen F., Multigrid Methods
Vol. 4	Coleman, Thomas F. and Van Loan, Charles, Handbook for Matrix Computations
Vol. 5	Grossman, Robert, Symbolic Computation: *Applications to Scientific Computing*
Vol. 6	McCormick, Stephen F., Multilevel Adaptive Methods for Partial Differential Equations
Vol. 7	Bank, R. E., PLTMG: A Software Package for Solving Elliptic Partial Differential Equations. *Users' Guide 6.0*
Vol. 8	Castillo, José E., Mathematical Aspects of Numerical Grid Generation
Vol. 9	Van Huffel, Sabine and Vandewalle, Joos, The Total Least Squares Problem: *Computational Aspects and Analysis*
Vol. 10	Van Loan, Charles, Computational Frameworks for the Fast Fourier Transform
Vol. 11	Banks, H.T. , Control and Estimation in Distributed Parameter Systems
Vol. 12	Cook, L. Pamela, Transonic Aerodynamics: *Problems in Asymptotic Theory*

Transonic Aerodynamics
Problems in Asymptotic Theory

Edited by L. Pamela Cook
University of Delaware

Society for Industrial and Applied Mathematics
siam
Philadelphia 1993

Library of Congress Cataloging-in-Publication Data

Transonic aerodynamics : problems in asymptotic theory / edited by L. Pamela Cook.
p. cm. — (Frontiers in applied mathematics ; 12)
Includes bibliographical references and index.
ISBN 0-89871-310-2
1. Aerodynamics, Transonic—Mathematics. 2. Asymptotic expansions. I. Cook, L. Pamela. II. Series.
TL571.T68 1993
629.132'304—dc20 93-3092

Cover art courtesy of Gilberto Schleiniger.

Contents

Contributors

L. Pamela Cook, Department of Mathematical Sciences, University of Delaware, Newark, Delaware 19716

N. D. Malmuth, Rockwell International Science Center, Box 1085, Thousand Oaks, California 91358-0085

Z. Rusak, Department of Mechanical Engineering, Aeronautical Engineering, and Mechanics, Rensselaer Polytechnic Institute, Troy, New York 12180-3590

Oleg S. Ryzhov, Department of Mathematical Sciences, Rensselaer Polytechnic Institute, Troy, New York 12181. Permanent address: Computing Center for the Russian Academy of Sciences, 40 Vavilov Street, Moscow 117333, Russia

Gilberto Schleiniger, Department of Mathematical Sciences, University of Delaware, Newark, Delaware 19716

Preface

Transonic aerodynamics involves flows for which the flow speed q is close to the local speed of sound a. That is, the Mach number $M = q/a \approx 1$. In this regime the flow is dominated by nonlinear effects. Present day commercial aircraft cruise in the transonic range. For an airplane flying in a uniform stream the drag rises precipitously as the Mach number far upstream approaches one. Thus there is a great interest in analyzing the transonic range of flight, its stability properties, and especially the question of designing reduced drag (shockless or weak shock) airfoils.

As pointed out above, the equations governing transonic flow are inherently nonlinear and must ultimately be solved numerically. However, asymptotic analysis has greatly simplified and enriched both the theory and the computations of these flows. Asymptotic analysis reduces the number of parameters involved, can simplify the geometry, and clarifies near and far field conditions. The asymptotic theory of transonic aerodynamics forms the basis for this monograph; it does not focus on computational aspects of transonic aerodynamics.

This monograph originated from a minisymposium organized by the editor at the second ICIAM meeting (International Conference on Industrial and Applied Mathematics) in Washington, D.C., July 1991. The purpose of the minisymposium was to gather researchers to talk on recent results in asymptotic and combined asymptotic/numeric treatment of problems arising in transonic aerodynamics. This resulting collection offers exciting results, perspectives, and case studies. It should be accessible to and of interest to any reader with a background and interest in aerodynamics, and perhaps also to the applied mathematician with a fluids/continuum mechanics background and an interest in nonlinear phenomena.

The monograph is organized into five chapters. The first chapter gives an introduction to transonic aerodynamics and to transonic small disturbance theory (TSD). The reader is provided with some background into the difficulties and inherent nonlinearity of transonic theory. The TSD limit is useful for thin bodies flying near the speed of sound in that it strips extraneous terms from the governing equations so that the simplest canonical description results. This chapter also provides an overview of the other chapters.

The second chapter, contributed by Ruzak, investigates the singularity in TSD due to the stagnation at the nose of a stationary airfoil in a windstream. Asymptotic expansions are used to investigate the structure of this nose region and to clarify the differences in the location of the stagnation point for subsonic and transonic flow.

In the TSD formulation viscosity is neglected, the rational being that it is important only in relatively thin layers. In Chapter 3, contributed by Ryzhov, the question of how the viscosity in these thin layers develops and affects the main flow and its stability is examined. He shows how to extend triple-deck theory of subsonic or supersonic boundary layers to viscous transonic flow. The most striking feature of his analysis is the importance of accounting for the normal pressure gradient, in contrast to classical Prandtl theory in which the normal pressure gradient plays no role.

Chapter 4, by Cook and Schleiniger, reviews results for solid wall choked wind-tunnel flows. Of interest here is the significant increase in interference as the flow approaches choking.

In Chapter 5, contributed by Malmuth, several problems in combined asymptotics and numerics are treated including the transonic area rule, wind-tunnel wall interference on a slender body, and a related problem in lubrication theory for foil bearings.

The asymptotic results set forth in this monograph exhibit some of the range of results available in asymptotics and in transonic flow. These asymptotic results yield similarity parameters, self-similar solutions and basic governing rules, and qualitative results that clarify and simplify the problems as well as the computational procedures.

I would like to express appreciation to all the contributors for their informative talks and articles and a special thank you to G. Schleiniger whose collaboration was crucial in completing this monograph. The influence of Julian D. Cole on this monograph should also be acknowledged. He is, in some sense, the father of this volume.

L. Pamela Cook
University of Delaware

Chapter 1

Introduction and Background: An Overview

L. Pamela Cook and Gilberto Schleiniger

Transonic aerodynamics involves flows about surfaces for which the flow speed q is close to the speed of sound a (the Mach number of the flow $M = q/a \approx 1$). These flows are dominated by nonlinear effects and have, in steady flow, both subsonic and supersonic regions within the flow. The flow is thus governed by a system of equations that, in steady flow, is of mixed type. The equations are nonlinear and they must ultimately be solved numerically. However, asymptotic analysis has greatly simplified and enriched both the theory and the computations of these flows. Asymptotic analysis reduces the number of parameters involved, can simplify the geometry, and clarifies near and far field conditions. It is the asymptotic theory of transonic aerodynamics which forms the basis for this monograph; this monograph does not focus on computational aspects of transonic aerodynamics. Much progress has been made during the last 20 years on computation of flows in the transonic regime, and surveys as well as research articles can be found elsewhere.

Present day commercial aircraft (e.g., the 747) cruise in the transonic range. For an airplane flying in a uniform stream, as the Mach number $M_\infty = U/a_\infty$ far upstream approaches one, the drag rises precipitously. Thus there is a great interest in analyzing the transonic range of flight, its stability properties, and especially the question of designing reduced drag (shockless or weak shock) wings.

One of the useful advances in transonic aerodynamics, valid for small disturbances (e.g., flow past thin airfoils or slender bodies) and in fact used throughout this monograph, is transonic small disturbance theory (TSD). This is a limiting formulation (valid as $M_\infty \to 1$ and the thickness ratio $\delta \to 0$). It is useful in that it strips extraneous terms from the governing equations so that the simplest canonical description results. In this monograph both inviscid and viscous transonic flows as well as steady and time-dependent flows are discussed. A few comments on two-dimensional (2D) steady inviscid flows are given below to orient the reader. For further background and more general flows see [1], [2] and the appropriate chapters in this monograph.

The basic equations governing steady, 2D, inviscid flow are conservation of mass

$$\text{div}(\rho\,\mathbf{q}) = 0\,, \tag{1.1}$$

balance of momentum

$$\text{div}(\rho\,\mathbf{q}\mathbf{q} + p\mathbf{I}) = \mathbf{0}\,, \tag{1.2}$$

and conservation of energy

$$\frac{\partial}{\partial x}\left\{\left(\frac{1}{2}\rho\,q^2 + \frac{p}{\gamma-1}\right) q_x + pq_x\right\} + \frac{\partial}{\partial y}\left\{\left(\frac{1}{2}\rho\,q^2 + \frac{p}{\gamma-1}\right) q_y + pq_y\right\} = 0\,, \tag{1.3}$$

where q_x, q_y are the components of the velocity vector $\mathbf{q}$, $q^2 = q_x^2 + q_y^2$, $\mathbf{I}$ is the identity (second-order tensor), $\gamma = c_p/c_v$ = ratio of specific heats (= 7/5 for a diatomic gas), p is the pressure, and ρ is the density. Associated with these equations are their integrated versions (conservation laws), which must hold across shocks. From these conservation laws, shock jump conditions are derived, namely

$$[\![\rho q_x]\!] dy_s - [\![\rho q_y]\!] dx_s = 0\,, \tag{1.4}$$

$$[\![\rho q_x^2 + p]\!] dy_s - [\![\rho q_y q_x]\!] dx_s = 0\,, \tag{1.5}$$

$$[\![\rho q_x q_y]\!] dy_s - [\![\rho q_y^2 + p]\!] dx_s = 0\,, \tag{1.6}$$

$$\left[\!\!\left[\left(\frac{1}{2}\rho q^2 + \frac{p}{\gamma-1}\right) q_x + pq_x\right]\!\!\right] dy_s - \left[\!\!\left[\left(\frac{1}{2}\rho q^2 + \frac{p}{\gamma-1}\right) q_y + pq_y\right]\!\!\right] dx_s = 0\,, \tag{1.7}$$

where $[\![\ \ ]\!] = (\ \ )_b - (\ \ )_a$ and $(\ \ )_{b,a}$ are the values behind and ahead of the shock, respectively.

By manipulation of these equations it can be shown that the enthalpy is constant on a streamline and hence, if a uniform state exists, for example far upstream, then

$$\frac{q^2}{2} + \frac{a^2}{\gamma-1} = \frac{U^2}{2} + \frac{a_\infty^2}{\gamma-1}, \quad a_\infty^2 = \frac{\gamma p_\infty}{\rho_\infty}\,;$$

that is, enthalpy is constant throughout the flow. It can also be shown that if ε measures the shock strength $\varepsilon = -[\![q]\!]/q_a$, then to order lower than ε^3, the entropy, and hence the vorticity, is constant. Thus, for weak shocks, to order lower than ε^3, a potential $\Phi(x,y)$ exists so that $\mathbf{q} = \nabla\Phi$. The system (1.1) – (1.3) becomes the *full potential* equation

$$(a^2 - \Phi_x^2)\,\Phi_{xx} - 2\Phi_x\Phi_y\Phi_{xy} + (a^2 - \Phi_y^2)\,\Phi_{yy} = 0\,, \tag{1.8}$$

with

$$\frac{1}{2}\left(\Phi_x^2+\Phi_y^2\right)+\frac{a^2}{\gamma-1}=\frac{1}{2}U^2+\frac{a_\infty^2}{\gamma-1}\,. \tag{1.9}$$

This description is an exact inviscid description if there are no shocks, e.g., in the subsonic range, and is accurate to order ε^2 if there are weak shocks, as for a small disturbance. Note that this system is nonlinear and may be of mixed (elliptic/hyperbolic) type. For a specific airfoil of chord 1, thickness δ, given by $y = \delta F_{u,\ell}(x)$ for $0 < x < 1$, where the subscripts u, ℓ indicate the upper and lower airfoil surfaces, appended to these equations must be boundary conditions:
(i) Uniform flow far upstream:

$$\Phi_x \to U\,,\ \Phi_y \to 0 \quad \text{as} \quad x \to -\infty\,, \tag{1.10}$$

(ii) Tangent flow at the airfoil surface $y = \delta F_{u,\ell}(x)$:

$$\frac{\Phi_y\left(x,\delta F_{u,\ell}(x)\right)}{\Phi_x\left(x,\delta F_{u,\ell}(x)\right)}=\delta F'_{u,\ell}(x)\,, \tag{1.11}$$

(iii) The Kutta-Joukowski condition,
(iv) the shock jump conditions (1.4) – (1.7), and
(v) the condition that entropy must increase across shock waves.

Traditional linear theory simplifies the boundary value problem immensely, in fact too much for transonic situations. Linear small disturbance theory assumes

$$\Phi\left(x,y;\,M_\infty,\delta\right)=U\left\{x+\delta\phi_1\left(x,y,M_\infty\right)+\delta^2\phi_2\left(x,y,M_\infty\right)+\cdots\right\}. \tag{1.12}$$

Substitution into the boundary value problem shows that ϕ_1 satisfies

$$\left(1-M_\infty^2\right)\phi_{1xx}+\phi_{1yy}=0\,, \tag{1.13}$$

$$\phi_{1x}\to 0 \quad \text{as} \quad x\to-\infty\,, \tag{1.14}$$

$$\phi_{1y}(x,0\pm)=F'_{u,\ell}(x)\,, \tag{1.15}$$

and the Kutta-Joukowski condition.

It is fairly simple to solve (1.13) – (1.15) for the supersonic ($M_\infty > 1$) flow past an object. In this case the equation is hyperbolic (a one-dimensional (1D) linear wave equation) and we get for $y > 0$

$$\phi_1=\begin{cases}0\,, & \text{for} \quad x-\sqrt{M_\infty^2-1}\,y\,<\,0;\\ \dfrac{-F_u\left(x-\sqrt{M_\infty^2-1}\,y\right)}{\sqrt{M_\infty^2-1}}\,, & \text{for} \quad 0\,<\,x-\sqrt{M_\infty^2-1}\,y\,<\,1;\\ \text{constant}\,, & \text{for} \quad 1\,<\,x-\sqrt{M_\infty^2-1}\,y\,.\end{cases}$$

In particular note that as $M_\infty \to 1$, ϕ_1 becomes singular. Further analysis [1] shows that

$$\phi_2 = O\left(\frac{1}{\left(M_\infty^2 - 1\right)^2}\right).$$

Clearly the expansion (1.12) is nonuniform and must be reformulated if

$$\frac{\delta^2\phi_2}{\delta\phi_1} = O\left(\frac{\delta}{\left(M_\infty^2 - 1\right)^{3/2}}\right) = O(1)\,.$$

Thus linearized theory is not valid if $M_\infty \to 1$, specifically if

$$1 - M_\infty^2 = O\left(\delta^{2/3}\right). \tag{1.16}$$

This clearly shows that linear small disturbance theory is no longer valid and a nonlinear small disturbance theory must be used for the transonic range. That analysis is reviewed below.

Transonic small disturbance theory treats the range for which linear theory breaks down, hence we keep the transonic similarity parameter

$$\frac{K = (1 - M_\infty^2)}{\delta^{2/3}}$$

fixed as $\delta \to 0$ to ensure we are in that range. Then

$$\Phi\left(x, y;\ M_\infty,\ \delta\right) = U\left\{x + \delta^{2/3}\phi\left(x, \tilde{y};\ K\right) + \cdots\right\},$$

where the expansion is valid as $\delta \to 0$ with K, x and $\tilde{y} = \delta^{1/3}y$ fixed.

This scaling was arrived at by attempting to find all significant limits of the full potential equation consistent with the boundary value problem [1]. In particular, since the airfoil is traveling near the speed of sound, disturbances travel with it and hence pile up. This means that disturbances occur on a faster scale in the direction of motion, the x direction, then in the transverse, y direction. In order to bring the disturbances onto the same scale, and preserve the body, the y coordinate must be stretched. The amount of stretching, $\delta^{-1/3}$, as well as the order of the potential disturbance, $\delta^{2/3}$, are determined by searching for significant limits of the boundary value problem. The resulting boundary value problem for ϕ is

$$\left(K - (\gamma + 1)\,\phi_x\right)\phi_{xx} + \phi_{\tilde{y}\tilde{y}} = 0\,, \tag{1.17}$$

$$\phi_x\,,\ \phi_{\tilde{y}} \to 0 \quad \text{as} \quad x \to -\infty\,, \tag{1.18}$$

$$\phi_{\tilde{y}}(x, 0\pm) = F'_{u,\ell}(x)\,, \quad 0 < x < 1\,, \tag{1.19}$$

(1.20) $$[\![\phi_x]\!]_{\mathrm{TE}} = 0 \quad \text{(the Kutta-Joukowski condition).}$$

In (1.20) the subscript TE indicates the trailing edge of the profile.

The boundary value problem (1.17) – (1.20) is the (first-order) TSD problem. The equation clearly changes from elliptic to hyperbolic as the values of ϕ_x increase from below to above $K/(\gamma+1)$. In this regime the limiting system has retained all necessary terms of the full potential equation and dispensed with the unnecessary terms. This formulation has a clear advantage over the full potential formulation in that the TSD equation is in conservation form. With $K = 0$ it admits a generalized two-parameter group of similarity transformations and thereafter the existence of self-similar solutions is established. The similarity parameter K, which appears only in the small disturbance theory, is a crucial ingredient in many conclusions relevant to engineering design.

The appended shock jump conditions are

(1.21) $$\left[\!\!\left[K\phi_x - \frac{\gamma+1}{2}\phi_x^2\right]\!\!\right] d\tilde{y}_s - [\![\phi_{\tilde{y}}]\!] dx_s = 0$$

and

(1.22) $$[\![\phi]\!] = 0\,.$$

Note that, as expected from linear theory blow up, the disturbance due to a body of small thickness ratio δ is much stronger, $O(\delta^{2/3})$, in the transonic range than in the subsonic or supersonic range where it is $O(\delta)$.

Numerous questions are still unanswered about either the full potential formulation or the TSD formulation, e.g., do solutions exist? Solutions have been found by various numerical schemes but have not been mathematically proven to exist. Are the solutions unique? How do the solutions depend on the Mach number? Do smooth (shock free) solutions exist and, if they do, what happens to them under slight perturbations? What are appropriate numerical schemes to solve these equations? These fascinating questions are not all addressed in this monograph. Here we are concerned with the validity of the TSD formulation, its limitations, and the properties of its solutions.

Since there is stagnation at the nose of a stationary airfoil in a windstream, any perturbation scheme based on a uniform flow will have a singularity at the nose. Chapter 2, by Rusak, investigates the singularity for a thin parabolic nose airfoil at a small angle of attack. Both subsonic and transonic flows are examined and the results contrasted. Asymptotic expansions of the velocity potential are constructed in terms of the thickness ratio of the airfoil in an outer region, around the airfoil, and in an inner region, near the nose, and matched. The outer expansion is the small disturbance theory where leading edge singularities appear; the inner expansion accounts for the stagnation flow at the nose. Uniformly valid expansions of the pressure distribution over the entire airfoil are found. This work demonstrates how useful the self-similar solutions are in studying the nonlinear problems inherent in transonic aerodynamics. Clear differences in the location of the stagnation point at the nose, and the symmetry of the pressure distribution, near the nose, for subsonic and transonic flows are indicated.

Viscosity is neglected in the basic TSD formulation, but real fluids are viscous, even though the effects of viscosity may be important only in relatively thin layers. Hence, there is the question of how the viscous flow in these thin layers develops and affects the main flow and its stability leading ultimately to laminar-turbulent transition. This question is considered in Chapter 3, by Ryzhov, where he shows how to extend the triple-deck theory of subsonic or supersonic boundary layer to viscous transonic flow. He distinguishes two different regimes: $M_\infty^2 - 1 = O(R^{-8/5})$ and $M_\infty^2 - 1 = O(R^{-8/9})$, where M_∞ is the free stream Mach number and R is the Reynolds number based on the free stream speed and the body characteristic length. In the first case, the perturbation potential obeys a nonlinear second-order equation (which coincides with (1.17) and is of mixed type in steady flow), whereas the Prandtl viscous wall sublayer equations are time-independent. In the second case, the perturbation potential satisfies a linear equation and the Prandtl equations are time-dependent. Here the unsteady interactions of pressure waves in the outer flow and vorticity oscillations in the viscous sublayer become a distinctive feature of the flow.

The classical triple-deck theory predicts a linearly stable boundary layer in supersonic regimes (not in accordance with actual measurements) and gives a dispersion curve that can lead to unstable oscillations in the subsonic regime in line with the Orr-Sommerfeld predictions. Ryzhov shows how to modify the classical theory to study the stability of Blasius boundary layer on a thermally insulated flat plate in the transonic range, from subsonic to moderately supersonic velocities. The most striking feature of his analysis is the importance of accounting for the normal pressure gradient when the frequency and wavenumber of an unstable eigenmode become large enough. This is in contrast to the classical Prandtl theory in which the normal pressure gradient plays no role whatsoever. Thus the contradiction between results of asymptotic analysis and experimental evidence that challenged theoreticians for more than 20 years may be considered to be settled now.

Chapter 4, by Cook and Schleiniger, is a review of results for choked solid wall wind-tunnel flows. Such flows are important limiting cases, hence in understanding wind-tunnel wall interference, that is, the effect the wind-tunnel walls have on the flow. Wind-tunnel wall interference calculations are crucial since models are tested in wind-tunnels. Both 2D and 3D, as well as lifting and nonlifting choked flows are discussed. Results of asymptotic theory predict the order of corrections to sonic free flow and the order of dependence of the choking Mach number on the wind-tunnel height. The interference increases significantly as the flow approaches choking as is discussed in this chapter. Numerical calculations are needed to confirm the asymptotic results and to obtain the actual corrections and numerical dependencies.

Finally Chapter 5, by Malmuth, considers several problems in combined asymptotics and numerics. The problems considered include the transonic area rule which, under appropriate conditions, shows that a 3D body has the same effect as an equivalent axisymmetric body. The consequent reduction of dimension by one significantly simplifies the numerical calculations. This result is useful in the second example where wind-tunnel wall interference on a slender body is treated. As noted above, analysis of the wall interference is crucial to design of airfoils. Lastly, asymptotics is used to simplify and clarify a problem of lubrication theory for foil bearings.

All of the chapters are tied together in that they illustrate the range of results possible in transonic aerodynamics through asymptotic and combined asymptotic/numeric treatment. Computational aspects of transonic aerodynamics are not discussed here. Computations can serve two purposes: to get numerical results for certain specific conditions and geometries, or to check or test analytical results or ideas. The focus of the articles in this monograph is instead on asymptotic analysis which provides understanding of the relevant phenomena for a broad range of conditions and geometries and also simplifies the nonlinear boundary value problems so that they yield more easily to modern computational methods. This approach is typical of modeling processes since all models are derived assuming certain parameters measuring contributions of various processes are small so that only *relevant* processes need be considered. The Euler equations, for example, are a limiting form (as viscosity vanishes) of the Navier-Stokes equations, which in turn are a limiting form of a more complete set of equations. The asymptotic limits set forth here, involving the TSD equation and the transonic boundary layer, are yet other steps in the process. These asymptotic results yield similarity parameters, self-similar solutions, basic rules (often used for design: the equal area rule, the stabilization law), and qualitative results that clarify and simplify the governing boundary value problems as well as the computational procedures.

References

[1] J. D. Cole and L. P. Cook, *Transonic Aerodynamics*, North Holland, Amsterdam, 1986.

[2] K. G. Guderley, *The Theory of Transonic Flow*, Pergamon Press, Oxford, 1962.

Chapter 2

Subsonic and Transonic Flows around a Thin Airfoil with a Parabolic Nose

Z. Rusak

2.1. Introduction

The subsonic and transonic flows about the leading edge of a thin airfoil with a round nose are complicated mathematical problems that also cause principal difficulties in the numerical solutions of these flows around the entire airfoil. The flow in the leading-edge region is characterized mainly by a stagnation point near the edge of the airfoil and also by very large gradients in the velocity and the pressure fields, leading to large discretization errors in the numerical calculations of the flow. Another difficulty that arises in the numerical solutions of the Euler equations around the nose is an entropy production that should not occur [6]. Therefore, a large number of mesh points and special refinement algorithms and many iterations are needed in order to calculate accurately the flows around the nose of the airfoil [6], [2]. The analysis of the flow in the leading-edge region can clarify the basic character of the flow in this region and can be helpful in the modern calculations of subsonic and transonic flows about thin airfoils with a round nose.

The analytical studies of the flow in the nose region, both in the subsonic regime and in the transonic regime, were based mainly on the small-perturbation potential flow theories. An approximation of the velocity potential function of the flow by the classical subsonic linearized theory [7] or by the transonic small-disturbance theory (TSD) [4] predicts infinite velocities and pressures near the leading edge of the airfoil, known as the subsonic—or transonic—"nose singularities." The pressure distribution changes like $(1/\sqrt{x})$ in the subsonic flow and like $(1/x^{1/3})$ in the transonic flow as $x \to 0$ (toward the leading edge of the airfoil), whereas actually the flow in the leading-edge region is brought continuously to a stagnation near the edge point. Also, comparisons between solutions of the small-disturbance problems and numerical solutions of the full potential-flow equations for two-dimensional (2D) airfoils indicated relatively large discrepancies in the nose region of the airfoil [16], [8]. This failure of the small-disturbance theories is due to the large perturbations to the free steam flow that occur in the leading-edge region and that must be calculated by a more exact theory.

The incompressible flow about a round-nosed thin airfoil was analyzed by Van Dyke [15], [16]. Asymptotic expansions of the surface speed, in an outer linearized flow around the airfoil's leading edge and in an inner flow around a parabola, were matched. The extension of this approach to subsonic flows is much more complicated due to the nonlinear compressibility effects and the absence of an exact solution to a subsonic flow around a parabola. Van Dyke [15] used the Janzen-Rayleigh expansion of a compressible flow past a parabola [5] and by a matching process derived a uniformly valid approximation of the surface speed along a thin airfoil. However, this approach is limited to small Mach numbers since Imai's expansion was carried up the M_∞^4-term only and further terms of this expansion were not derived.

The theoretical study of the transonic flow around the leading edge of a thin airfoil with a parabolic nose was limited to the analysis of the transonic small-disturbance nose-singularity only [8], [4]. Keyfitz, Melnik, and Grossman [8] described the small-disturbance flow around the leading edge by an asymptotic series of similarity terms that were given by a set of boundary value problems and were solved numerically. It was found that the flow near the nose is dominated by the thickness effects, where the leading term corresponds to a sonic symmetric flow over a slender parabola at zero incidence. In the general case, the next possible term after the leading symmetric term corresponds to an eigenfunction of the problem that represents an antisymmetric flow around the nose, with an eigenvalue exponent that is very close to the leading term exponent. On the other hand, Cole and Cook [4] tried to approximate the dominant term of the perturbation velocity-potential near the nose by using hodograph-similarity solutions of the sonic small-disturbance problem. It was assumed that the hodograph singular solution near the nose is composed of two symmetric and asymmetric terms of the same leading order to represent the thickness and circulation effects around the nose. However, as was shown in Rusak [11], this solution is inconsistent with the satisfaction of the boundary condition on the parabolic nose. This was corrected and analytical expressions that coincide with the numerical solutions of Keyfitz, Melnik, and Grossman [8] were found.

Despite being two important problems, both theoretically and numerically, the analyses of the subsonic and the transonic potential flows about the leading edge of a thin round-nosed airfoil were never completed. Only recently have two consistent analyses been presented by Rusak [10], [11] for the solutions of these problems. Matched asymptotic methods were used to analyze the problems. Uniformly valid pressure distributions on the entire airfoil surface were derived for an airfoil given in a subsonic and in a transonic flow at a small angle of attack. The analyses show that to the orders discussed, in a subsonic flow ($0 \leq M_\infty < M_{\text{cr}}$) the stagnation point tends to shift toward the leading edge of the airfoil due to the compressibility effects as the Mach number

(M_∞) of the oncoming flow is increased, whereas in a transonic flow ($M_\infty \sim 1$) the stagnation point is located at the leading edge of the airfoil (at the most forward point of the airfoil against the oncoming flow) for every transonic Mach number and shape and angle of attack of the airfoil. The analyses also show that the pressure distributions around the nose change differently in the two cases. In a subsonic flow the pressure distribution around the nose is strongly affected from the total circulation around the airfoil. However, in a transonic flow the pressures on the upper and lower surfaces are locally symmetric near the edge point, and asymmetric deviations increase and become significant only when the distance from the leading edge of the airfoil increases beyond the edge region.

Presented in this paper is an overview of the recent two analyses of Rusak [10], [11]. The results of these analyses are then discussed and a possible physical mechanism is suggested to explain the basic differences between a subsonic flow and a transonic flow that appear in the leading-edge region of a thin airfoil with a parabolic nose. The results of these analyses were recently applied to calculate the pressure distributions around the leading edge of several airfoils given at various Mach numbers and angles of attack. A good agreement was found in the leading-edge region between the present solutions and refined numerical solutions of the full potential-flow equations and of the Euler equations [14].

2.2. Basic Problem and Equations

A subsonic or transonic potential flow about a 2D thin airfoil with a parabolic nose given at a small angle of attack is considered in a (x, y) plane with unit vectors $(\vec{e}_x, \vec{e}_y)$ (Fig. 2.1). The airfoil shape is given by

$$B(x,y) = y - \delta F_{u,\ell}(x) = 0\,, \quad 0 \le x \le c\,, \tag{2.1}$$

where c is the airfoil's chord and δ is the thickness ratio, $\delta \ll 1$. The functions $F_{u,\ell}(x)$ represent the upper and lower surfaces, respectively. These shape-functions are described by

$$F_{u,\ell}(x) = C_a(x) - Ax \pm ct\left(\frac{x}{c}\right) \quad \text{for} \quad 0 \le x \le c\,, \tag{2.2}$$

where $C_a(x)$ is the camber line function, $A = \alpha/\delta, \alpha$ is the angle of attack, and $ct(x/c)$ is the thickness distribution function. Also, $t(0) = t(1) = 0$ and $C_a(0) = C_a(c) = 0$. Near the leading edge as $x \to 0$ the thickness function changes like $ct(x/c) \sim 2h\sqrt{cx} + O(x)$ and the camber function changes like $C_a(x) \sim a_1 x + O(x^q)$ with $(q > 1)$, where $R_c = 2h^2\delta^2 c$ is the radius of curvature of the parabolic nose, and (a_1) is the local camber of the airfoil at the leading edge (see [1]).

To the orders considered both the subsonic and transonic flows are irrotational and isentropic. The velocity-potential field Φ of the flow, where $\vec{q} = \nabla\Phi$

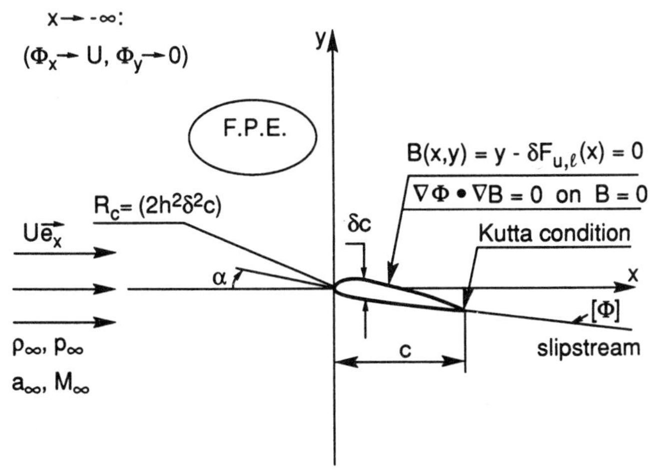

FIG. 2.1. *Airfoil problem.*

is the velocity vector, is described by the full potential-flow equation:

$$(2.3)\qquad \begin{gathered}(a^2-\Phi_x^2)\Phi_{xx}-2\Phi_x\Phi_y\Phi_{xy}+(a^2-\Phi_y^2)\Phi_{yy}=0\,,\\ \frac{a^2}{U^2}=\frac{1}{M_\infty^2}+\frac{\gamma-1}{2}\left[1-\frac{\Phi_x^2+\Phi_y^2}{U^2}\right],\quad M_\infty\equiv\frac{U}{a_\infty}\,,\end{gathered}$$

where U, a_∞, M_∞ are the speed, speed of sound, and Mach number of the flow at upstream infinity, respectively, (a) is the local speed of sound, and γ is the ratio of specific heats. The solution of (2.3) is governed by the tangency boundary condition on the airfoil surface,

$$(2.4)\qquad \nabla\Phi\cdot\nabla B=0 \quad \text{on} \quad B=0\,.$$

Also, disturbances must die out at upstream infinity as $x\to-\infty:(\Phi_x\to U,\Phi_y\to 0)$, and the Kutta condition is satisfied at a sharp subsonic trailing edge. In order to get a one-valued potential function, the (x, y) plane is considered as cut along the streamline, which leaves the trailing edge to infinity, where the potential Φ is allowed to jump due to the circulation around the airfoil (Fig. 2.1). Shock jump conditions, resulting from the conservation form of (2.3), should also be satisfied on any shock wave that arises from the solution, and only compression jumps are allowed. The density (ρ) and pressure (p) fields of the flow are calculated as functions of the velocity $\vec{q}$ by the isentropic relations and the conditions at upstream infinity $(\rho_\infty, p_\infty, \vec{q}=U\,\vec{e}_x)$:

$$(2.5)\qquad \frac{\rho}{\rho_\infty}=\left[1+\frac{\gamma-1}{2}M_\infty^2\left(1-\frac{|\,\vec{q}\,|^2}{U^2}\right)\right]^{\frac{1}{\gamma-1}},\quad \frac{p}{p_\infty}=\left[\frac{\rho}{\rho_\infty}\right]^\gamma.$$

In order to study the subsonic flow about a thin airfoil with a round nose, the potential function Φ is approximated by asymptotic expansions in the limit $\delta \to 0$ with M_∞ and A fixed. Outer expansions are constructed in an outer region around the airfoil, with fixed coordinates (x, y) as $\delta \to 0$. Inner expansions are constructed in the nose region using fixed stretched coordinates $(x^* = x/\delta^2,\ y^* = y/\delta^2)$ as $\delta \to 0$.

In order to study the transonic flow around the leading edge of a thin airfoil with a parabolic nose, the potential function Φ is approximated by asymptotic expansions in the limit $(\delta \to 0, M_\infty \to 1)$ with the transonic similarity parameter $K = (1 - M_\infty^2)/\delta^{2/3}$ and A fixed. Outer expansions are constructed in an outer region around the leading edge of the airfoil with fixed coordinates $(x, \tilde{y} = \delta^{1/3} y)$ as $\delta \to 0$. Inner expansions are constructed in the nose region using fixed stretched coordinates $(x^* = x/\delta^2, y^* = y/\delta^2)$ as $\delta \to 0$.

In each of the flow regimes, the outer expansion consists of the small-disturbance theory and its second-order problem where the nose singularity appears. The inner expansion accounts for the compressible flow around the nose of the airfoil where a stagnation point exists. In the inner region, the picture of the flow is magnified by $(1/\delta^2)$ in order to capture the basic character of the flows around the nose. The outer and inner expansions of each problem are matched in an intermediate region in between the scalings of the outer and the inner region. The outcome of the matching process is two well-defined boundary value problems for the solution of the inner compressible potential flow around an infinite 2D parabola surface. Using the outer small-disturbance solutions and the nose inner solutions, uniformly valid pressure distributions on the entire airfoil surface in a subsonic and in a transonic flow are derived from the composite solutions.

2.3. The Subsonic Problem

In the subsonic problem (where $0 \le M_\infty < M_{cr}$ and M_{cr} is the critical Mach number), the outer limit process of the first- and second-order airfoil theory has $\delta \to 0$ and (x, y, M_∞, A) fixed. The outer expansion has the form:

$$\Phi(x, y; M_\infty, \delta) = U\{x + \delta\phi_1(x, y; \beta, A) + \delta^2\phi_2(x, y; \beta, A) + O(\epsilon(\delta))\}, \tag{2.6}$$

where $\beta = \sqrt{1 - M_\infty^2}$ and $\epsilon(\delta) << \delta^2$. From the basic system of equations (2.1) to (2.5), a sequence of problems is found for the solution of the functions ϕ_1 and ϕ_2. ϕ_1 satisfies the subsonic linear theory equation and ϕ_2 a forced equation (Fig. 2.2):

$$\begin{aligned} &\beta^2\phi_{1xx} + \phi_{1yy} = 0, \\ &\phi_{1y}(x, 0\pm) = F'_{u,\ell}(x) \quad \text{for} \quad 0 \le x \le c, \\ &(\phi_{1x}, \phi_{1y}) \to 0 \quad \text{as} \quad x \to -\infty, \\ &\phi_{1x}(c, 0^+) = \phi_{1x}(c, 0^-), \end{aligned} \tag{2.7}$$

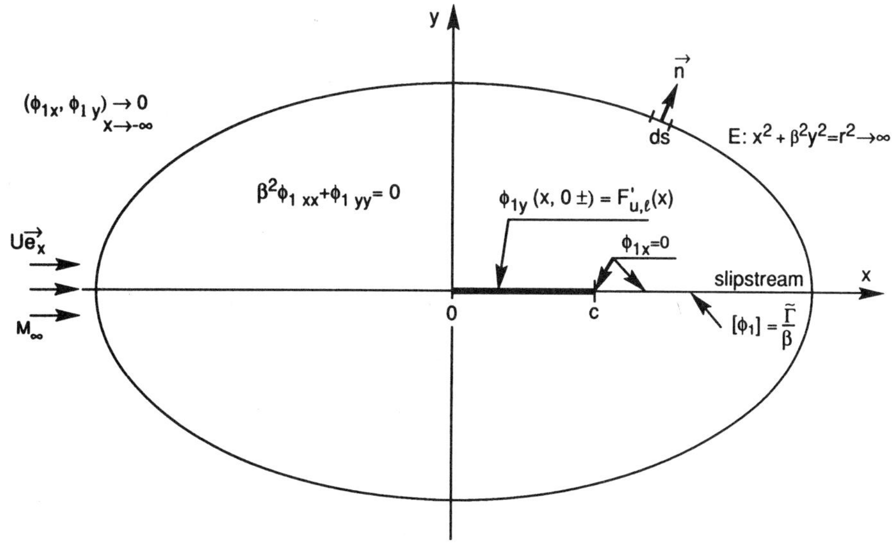

FIG. 2.2. *Linearized airfoil problem (outer region).*

$$
\begin{aligned}
&\beta^2\phi_{2xx} + \phi_{2yy} = 2M_\infty^2\left\{\left[1 + \frac{\gamma - 1}{2}M_\infty^2\right]\phi_{1x}\phi_{1xx} + \phi_{1y}\phi_{1xy}\right\}, \\
&\phi_{2y}(x, 0\pm) = \phi_{1x}(x, 0\pm)F'_{u,\ell}(x) - \phi_{1yy}(x, 0\pm)F_{u,\ell}(x) \text{ for } 0 \le x \le c, \\
&(\phi_{2x}, \phi_{2y}) \to 0 \text{ as } x \to -\infty.
\end{aligned} \tag{2.8}
$$

The solution of the function ϕ_1 is composed of a lifting part due to the angle of attack and the camber of the airfoil, and a symmetric part due to the airfoil thickness. The lifting part is described by a distribution of potential subsonic vortices along the x-axis segment $(0, c)$ and the symmetric part by a distribution of potential subsonic sources. The solution of the function ϕ_2 is found by using Van Dyke [16] solution of the second-order problem. As the leading edge of the airfoil is approached (as $r \to 0$ where $x = r\cos\theta, \beta y = r\sin\theta$) the solutions of (2.7) and (2.8) show that [10]:

(2.9)

$$
\begin{aligned}
\Phi \sim U\Bigg\{ & x + \delta\left[\frac{2h\sqrt{c}}{\beta}\sqrt{r}\sin\frac{\theta}{2} + \frac{2A\omega_0\sqrt{c}}{\beta}\sqrt{r}\cos\frac{\theta}{2} + \text{const.}\right. \\
& \left. + \frac{h_2}{\pi\beta}\left[r\log\left(\frac{r}{c}\right)\cos\theta + r(\pi - \theta)\sin\theta - r\cos\theta\right] + ux + vy + O\left[\frac{r^{3/2}}{\sqrt{c}}\right]\right] \\
& + \delta^2\left[\text{const.} + Q\log r + \frac{M_\infty^2 h^2 c}{\beta^2}\left[(m_1(\beta)\cos\theta + m_2(\beta)\cos 2\theta)\right.\right. \\
& + \frac{2A\omega_0}{h}m_2(\beta)\sin 2\theta + \frac{A^2\omega_0^2}{h^2}(m_1(\beta)\cos\theta - m_2(\beta)\cos 2\theta) \\
& \left.\left. + \frac{A\omega_0}{h}m_1(\beta)\theta\right] + O\left[\sqrt{r}\log\frac{r}{c}, \sqrt{r}\right]\right] + O(\epsilon(\delta))\Bigg\},
\end{aligned}
$$

where

$$(2.10)\qquad \begin{aligned} m_1(\beta) &= 1 - 4m_2(\beta), \quad m_2(\beta) = -\frac{\gamma+1}{16}\frac{M_\infty^2}{\beta^2}, \\ \omega_0 &= 1 - \frac{1}{\pi A}\int_0^\pi C_a'(\xi)d\mu, \quad \xi = \frac{c}{2}(1+\cos\mu),\ 0 \le \mu \le \pi, \end{aligned}$$

and h_2, u, v are known constants that are related to the given shape functions of the airfoil. Equation (2.9) shows that the outer small-disturbance flow around the nose is described in its leading term by both symmetric and circulation effects. In (2.9), the source strength Q of the second-order term is yet undetermined and its value will be determined by matching the outer and inner solutions. Also, the second-order term involves a concentrated point vortex term in order to satisfy the tangency boundary condition for ϕ_2 as $x \to 0$. It is a result of the combined effects of compressibility, angle of attack, camber, and the thickness of the airfoil. The circulation strength of this vortex depends on the Mach number (M_∞), and when $M_\infty = 0$ this vortex term vanishes.

Equation (2.9) shows that in the leading-edge region as $r \to 0$, the velocities both in the x- and y-directions become singular like $(\delta/\sqrt{r})$, specifically on the airfoil surface ($r = x, \theta = 0, 2\pi$). This is known as the subsonic "leading-edge singularity." There is also a misordering in the magnitude of the disturbances for every θ when r becomes smaller than δ^2. Therefore, a rescaling in the radial direction only is needed, $r^* = r/\delta^2$, in order to account for the local compressible flow around the airfoil's nose, where a stagnation point exists.

An inner limit process defines an expansion valid near the nose where $\delta \to 0$ and $(x^*, y^*; M_\infty, A)$ fixed and $x^* = x/\delta^2, y^* = y/\delta^2$. The form of the expansion is

$$(2.11)\qquad \Phi(x, y; M_\infty, \delta) = U\delta^2\phi_0(x^*, y^*; M_\infty, A).$$

Because of the uniform scaling, ϕ_0 satisfies the exact potential equation (2.3) in the (x^*, y^*) plane. ϕ_0 describes in the inner region a uniform flow with Mach number (M_∞) about a parabola surface where the tangency boundary condition has to be satisfied (Fig. 2.3):

$$(2.12)\qquad \phi_{0y^*} = \pm\frac{h\sqrt{c}}{\sqrt{x^*}}\phi_{0x^*} \text{ on } y^* = \pm 2h\sqrt{cx^*}, \quad 0 \le x^* < \infty.$$

The expansion of the solution to the inner problem shows that as $r^* =$

$\sqrt{x^{*2}+\beta^2 y^{*2}} \to \infty$ [10]:

(2.13)

$$\begin{aligned}\Phi \sim U\delta^2 \Bigg\{ & x^* + \frac{2h\sqrt{c}}{\beta}\sqrt{r^*}\sin\frac{\theta^*}{2} + \frac{2\bar{A}}{\beta}\sqrt{r^*}\cos\frac{\theta^*}{2} \\ & + \frac{M_\infty^2 h^2 c}{\beta^2}\Bigg[(m_1(\beta)\cos\theta^* + m_2(\beta)\cos 2\theta^*) \\ & + 2\frac{\bar{A}}{h\sqrt{c}}m_2(\beta)\sin 2\theta^* + \frac{\bar{A}^2}{h^2 c}(m_1(\beta)\cos\theta^* - m_2(\beta)\cos 2\theta^*) \\ & + \frac{\bar{A}}{h\sqrt{c}}m_1(\beta)\theta^*\Bigg] + \bar{Q}\log r^* + O\left[\frac{\log r^*}{\sqrt{r^*}}, \frac{1}{\sqrt{r^*}}\right]\Bigg\},\end{aligned}$$

where $\theta^* = \theta$, the vortex strength $\bar{A}$ is yet undetermined and, where the source term $\bar{Q}$ is included to satisfy the conservation of mass in the inner flow problem,

$$\bar{Q} = (\bar{A}^2 - h^2 c)\frac{M_\infty^2}{2\beta^2}m_1(\beta)\,. \tag{2.14}$$

The matching between the near field of the outer expansion and the far field of the inner expansion is carried out with the help of an intermediate limit $\delta \to 0$ with $r_\eta = r/\eta(\delta)$ fixed where $\delta^2 << \eta(\delta) << 1$ such that as $\delta \to 0, \eta(\delta)/\delta^2 \to \infty$. Then, $r = \eta(\delta)r_\eta \to 0$ and $r^* = \eta r_\eta/\delta^2 \to \infty$. $\eta(\delta)$ represents a whole order class of limit between the inner and the outer regions and is called the overlap region. For matching, the expansions must read the same to a certain order when expressed in the r_η coordinate. The outer and inner expansions match ((2.9) and (2.13)) if

$$\begin{aligned}\bar{A} &= A\omega_0\sqrt{c}\,, \\ Q = \bar{Q} &= \frac{M_\infty^2}{2\beta^2}m_1(\beta)(A^2\omega_0^2 - h^2)c\,.\end{aligned} \tag{2.15}$$

The matching process results in a well-defined boundary-value problem in the inner region for the solution of a compressible potential flow around an infinite parabola surface $y^* = \pm 2h\sqrt{cx^*}$ at zero angle of attack. The far-field expansion of the inner flow is composed of a uniform flow with the same Mach number (M_∞) as of the oncoming flow about the airfoil, and of a circulation flow due to the total circulation ($A\omega_0/h$) around the airfoil and to the compressibility effects (see (2.13)). The inner problem is solved numerically by using a transformation to parabolic coordinates and a finite-differences scheme with a point over-relaxation algorithm. The numerical calculations provide the pressure distribution C_p^* around the parabolic nose with a given Mach number (M_∞) and circulation parameter ($A\omega_0/h$). Examples of the calculated pressure distributions are shown in Fig. 2.4. Results show that the basic effect of the circulation is to move the stagnation point downstream along the lower surface. However, the stagnation point tends to shift upstream as the Mach

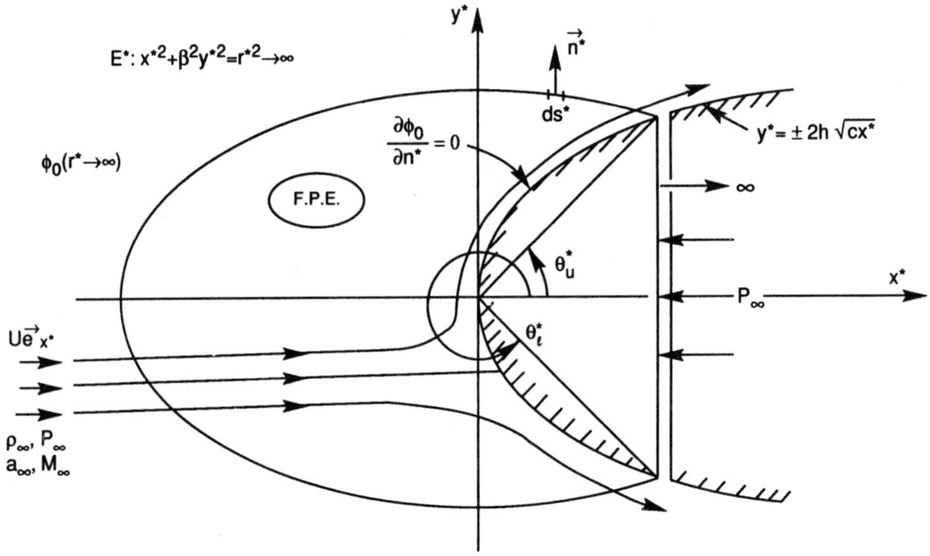

FIG. 2.3. *Parabolic nose problem (inner region).*

number is increased for a fixed value of the circulation (Fig. 2.5). This occurs because the total compressible effect is to reduce the circulation around the parabola nose due to the point vortex term in the far-field expansion of the inner flow that opposes the basic circulation terms (see (2.13)).

A uniformly valid solution for the potential Φ can be constructed from the solution of the outer linearized problem for ϕ_1 and the parabola inner solution for ϕ_0 by adding the two together and subtracting the common part in the intermediate overlap region where the two solutions match. The composite solution in the limit $\delta \to 0$ with M_∞, A fixed is given by

(2.16)
$$\Phi(x,y;M_\infty,A,\delta) \sim U\{x+\delta\phi_1(x,y;\beta,A)+\ldots+\delta^2\phi_0(x^*,y^*;M_\infty,A) - \ldots - \Phi_{\text{c.p.}}(r_\eta,\theta;\delta,\eta(\delta),M_\infty,A)\},$$

where

(2.17)
$$\begin{aligned}\Phi_{\text{c.p.}}(r_\eta,\theta;\delta,\eta(\delta),M_\infty,A) &= \eta x_\eta + \frac{2h\sqrt{c}}{\beta}\delta\sqrt{\eta}\sqrt{r_\eta}\cos\frac{\theta}{2} \\ &+ \frac{2A\omega_0\sqrt{c}}{\beta}\delta\sqrt{\eta}\sqrt{r_\eta}\cos\frac{\theta}{2} + \delta^2 Q\log(\eta r_\eta) \\ &+ \delta^2\frac{M_\infty^2 h^2 c}{\beta^2}\left[(m_1(\beta)\cos\theta + m_2(\beta)\cos 2\theta) + \frac{2A\omega_0}{h}m_2(\beta)\sin 2\theta\right. \\ &\left. + \frac{A^2\omega_0^2}{h^2}(m_1(\beta)\cos\theta - m_2(\beta)\cos 2\theta) + \frac{A\omega_0}{h}m_1(\beta)\theta\right].\end{aligned}$$

Using the composite solution and the isentropic relations, a uniformly valid pressure distribution over the entire airfoil is obtained. In the limit as $\delta \to 0$

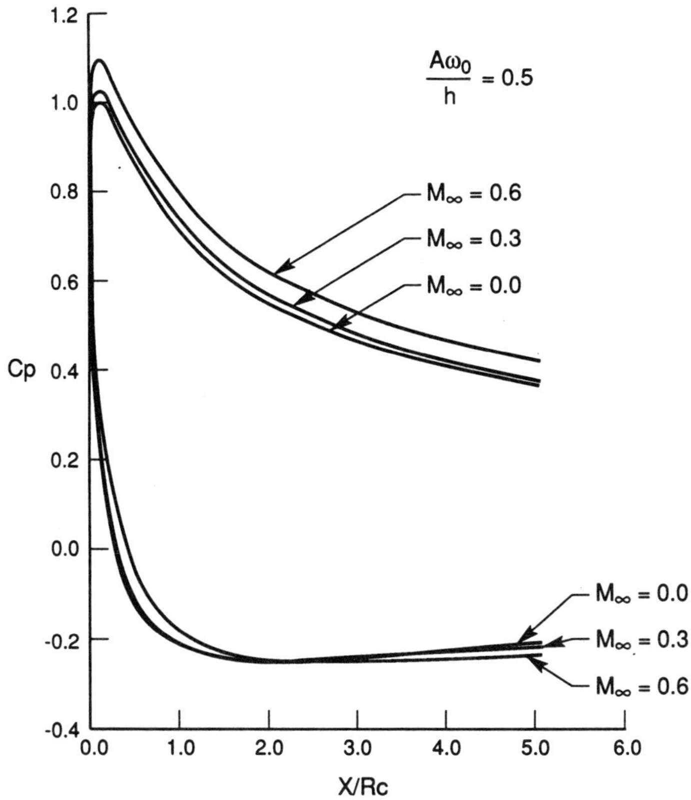

FIG. 2.4. *The pressure distribution along the parabolic nose for* $\frac{A\omega_0}{h} = 0.5$ *and various Mach numbers.*

with $(M_\infty, A\omega_0/h)$ fixed:

$$C_p(x; \delta, M_\infty, A) = C_p^*\left(\frac{x}{R_c}; M_\infty, \frac{A\omega_0}{h}\right) + \\ + \left[C_{p_{u,\ell}}(x; \beta, A) - C_{\text{pc.p.}}(x; \delta, M_\infty, A)\right]\frac{\rho^*}{\rho_\infty}\phi_{0x^*} \,. \tag{2.18}$$

Here $C_p^*(x/R_c; M_\infty, A\omega_0/h)$, ρ^*/ρ_∞, and ϕ_{0x^*} are calculated from the numerical solution of the subsonic parabola problem (inner problem), and $C_{p_{u,\ell}}(x; \beta, A)$ is the outer pressure coefficient on the upper and lower surfaces of the airfoil that is calculated by the subsonic linearized theory. $C_{\text{pc.p.}}$ is the common part of the pressure distribution

$$C_{\text{pc.p.}} = \mp \frac{2A\omega_0\delta}{\sqrt{1 - M_\infty^2}}\sqrt{\frac{c}{x}} \,. \tag{2.19}$$

It coincides with the classical leading-edge singularity of the linearized pressure distribution $(C_{p_{u,\ell}})$ due to the angle of attack and the camber of the airfoil.

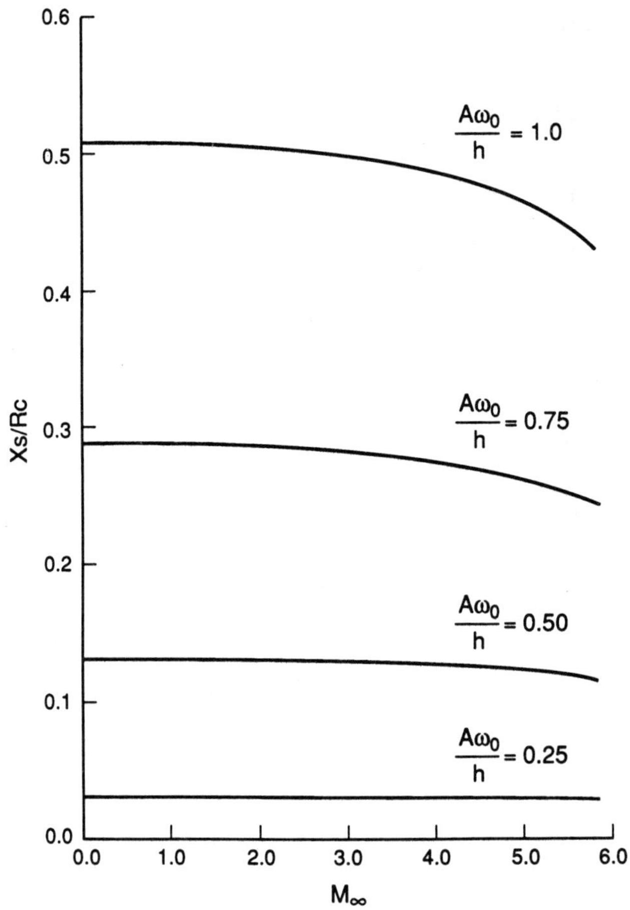

FIG. 2.5. *The shift of the stagnation point with Mach number and various circulation values.*

As the leading edge of the airfoil is approached ($x \to 0$ or $0 < x < R_c$), the common pressure coefficient $C_{\mathrm{pc.p.}}$ cancels the nose singularity of the outer pressure coefficient ($C_{p_{u,\ell}}$) both on the upper and lower surfaces of the airfoil. Also in this region (ϕ_{0x^*}) is small and tends to zero near the stagnation point. Therefore, the dominant term very close to the leading-edge point is the parabola pressure coefficient (C_p^*). As x is increased beyond the leading-edge region ($x > R_c/\delta$) the term $((\rho^*/\rho_\infty)\phi_{0x^*})$ tends to one and the common part $C_{\mathrm{pc.p.}}$ cancels the parabola pressure (C_p^*). Therefore, the dominant term outside the leading-edge region is the outer pressure coefficient ($C_{p_{u,\ell}}$) that is calculated by the linearized theory. In the intermediate region, the pressure distributions change uniformly from (C_p^*) to ($C_{p_{u,\ell}}$). Using this approach, the pressure distributions around the entire shape of a thin airfoil with a parabolic nose that is given in a subsonic flow and at a small angle of attack can be approximated.

2.4. The Transonic Problem

In the transonic regime (where $M_\infty \sim 1$), the outer limit process of the first- and second-order airfoil theory has $(\delta \to 0, M_\infty \to 1)$ where

$$\frac{1}{M_\infty^2} = 1 + K\delta^{2/3} + K_1\delta^{4/3} + \ldots$$

and $(x, \tilde{y} = \delta^{1/3}y, K, K_1, A)$ fixed. The outer expansion has the form

$$\begin{aligned}(2.20)\qquad &\Phi(x, y; M_\infty, \delta) \\ &= U\{x + \delta^{2/3}\phi_1(x, \tilde{y}; K, A) + \delta^{4/3}\phi_2(x, \tilde{y}; K, K_1, A) + O(\epsilon(\delta))\}\,,\end{aligned}$$

where $\epsilon(\delta) << \delta^{4/3}$. From the basic system of equations (2.1) to (2.5), a sequence of outer problems is found for the solution of the functions ϕ_1 and ϕ_2. The transonic small-disturbance problem for ϕ_1 is given by Cole and Cook [4] (Fig. 2.6):

$$(2.21)\qquad \begin{aligned}&(\gamma+1)\phi_{1x}\phi_{1xx} - \phi_{1\tilde{y}\tilde{y}} = K\phi_{1xx}\,,\\ &\phi_{1\tilde{y}}(x, 0\pm) = F'_{u,\ell}(x) \text{ for } 0 \le x \le c\,,\\ &(\phi_{1x}, \phi_{1\tilde{y}}) \to 0 \text{ as } x \to -\infty\,,\\ &\phi_{1x}(c, 0^+) - \phi_{1x}(c, 0^-) = 0\,.\end{aligned}$$

The second-order problem for ϕ_2 is

$$(2.22)\qquad \begin{aligned}&(K - (\gamma+1)\phi_{1x})\phi_{2xx} + \phi_{2\tilde{y}\tilde{y}} - (\gamma+1)\phi_{1xx}\phi_{2x}\\ &= -K_1\phi_{1xx} - K\phi_{1\tilde{y}\tilde{y}} + \frac{\gamma+1}{2}\phi_{1x}^2\phi_{1xx} + 2\phi_{1\tilde{y}}\phi_{1x\tilde{y}} + (\gamma-1)\phi_{1x}\phi_{1\tilde{y}\tilde{y}}\,,\\ &\phi_{2\tilde{y}}(x, 0\pm) = \phi_{1x}(x, 0\pm)F'_{u,\ell}(x) \text{ for } 0 \le x \le c\,,\\ &(\phi_{2x}, \phi_{2\tilde{y}}) \to 0 \text{ as } x \to -\infty\,.\end{aligned}$$

The solution of (2.21) is approximated in the nose region (as $x \to 0$) by a sum of similarity solutions:

$$(2.23)\qquad (\gamma+1)\phi_1 \sim \tilde{y}^m f(\xi) + \tilde{y}^a f_p(\xi) + \ldots, \quad \xi = \frac{x}{\tilde{y}^k}, \quad m < a\,,$$

where k, m, a are constants, ξ is the similarity variable, and f, f_p are the similarity functions. The substitution of (2.23) into (2.21) results for the case of a parabolic nose in [4]:

$$(2.24a)\qquad k = 6/7, \quad m = 4/7\,.$$

Therefore, to the leading term:

$$(2.24b)\qquad (\gamma+1)\phi_1 \sim \tilde{y}^{4/7} f(\xi) + \ldots, \quad \xi = \frac{x}{\tilde{y}^{6/7}}\,,$$

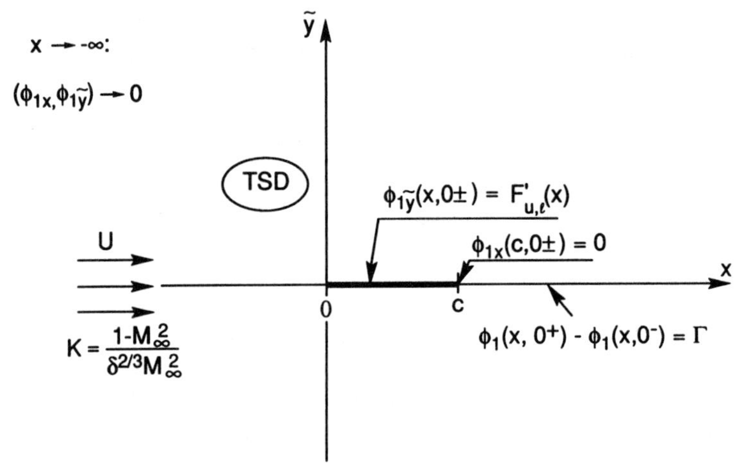

FIG. 2.6. *Transonic small-disturbance problem.*

where f is determined by the nonlinear equation:

$$\left[f_\xi - \frac{36}{49}\xi^2\right] f_{\xi\xi} - \frac{30}{49}\xi f_\xi + \frac{12}{49} f = 0\,. \tag{2.25}$$

The solution of (2.25) is found by using similarity solutions of the sonic small-disturbance equation that were recently presented by Rusak [12], [13]. These analytical solutions were derived from similarity solutions of the problem in the hodograph plane (ρ, ν) where

$$\begin{cases} \rho^2 = v^2 - \frac{4}{9}\omega^3, & \sin\nu = v/\rho\,, \\ \omega = (\gamma+1)\phi_{1x}, & v = (\gamma+1)\phi_{1\tilde{y}}\,. \end{cases} \tag{2.26}$$

In the case of a parabolic nose, the singular solution $\tilde{y}$ is given by Rusak [11] or Cole and Cook [4]:

$$\begin{aligned} &\tilde{y} = c_1 \rho^{-7/3} \mathcal{F}(\sin\nu)\,, \\ &\mathcal{F}(\sin\nu) \equiv c_2\left[1 - \frac{7}{3}\sin^2\nu\right] + \sin\nu F\left(\frac{5}{3} - \frac{1}{2}; \frac{3}{2}; \sin^2\nu\right), \end{aligned} \tag{2.27}$$

where c_1 and c_2 are constants to be determined, F is the standard hypergeometric function [3], and $\mathcal{F}$ is the solution of the hodograph-hypergeometric equation, given in its self-adjoint form by

$$\cos^2\nu \mathcal{F}'' - \frac{4}{3}\sin\nu \mathcal{F}' + \frac{14}{3}\mathcal{F} = 0, \quad \mathcal{F}' \equiv \frac{d\mathcal{F}}{d(\sin\nu)}\,. \tag{2.28}$$

The solution $\tilde{y}$ in (2.27) is described by a linear combination of a symmetric and unsymmetric functions, in terms of the hodograph similarity variable $(\sin\nu)$ to represent the symmetric and circulation effects around the nose of the airfoil as occurs in the subsonic case. The parameter ν changes in the range $\nu_1 \leq \nu \leq \nu_3$ where $\nu_1 < 0, \nu_2$ and $\nu_3 > 0$ are the first three roots around $\nu = 0$ of the equation $\tilde{y} = 0$ or $\mathcal{F}(\sin\nu) = 0$.

By using the transformation relations [9],

$$Z = 1 - \frac{9}{4}\frac{\sigma^2}{t^3}, \quad W = \frac{t^2}{\sigma}, \tag{2.29a}$$

where t and σ are the phase-plane parameters [4]

$$\omega = \left[\frac{x}{\tilde{y}}\right]^2 t, \quad v = \left[\frac{x}{\tilde{y}}\right]^3 \sigma, \tag{2.29b}$$

the hodograph solution of (2.27) can be transferred back to the physical plane to calculate the similarity function $f(\xi)$. The solution is given by a parametric representation in terms of the hodograph similarity variable ν [11]:

$$\begin{aligned} \xi &= -\frac{1}{2}\left[\frac{3}{2}\right]^{1/3} c_1^{1/7}\frac{\cos^{4/3}\nu\mathcal{F}'}{\mathcal{F}^{6/7}}, \\ f &= \frac{1}{8}c_1^{3/7}\left[14\mathcal{F}^{3/7}\sin\nu + 9\frac{\mathcal{F}'\cos^2\nu}{\mathcal{F}^{4/7}}\right]. \end{aligned} \tag{2.30}$$

The substitution of (2.24) and (2.30) into the boundary condition in (2.21) results in only one consistent solution of the constants, c_1 and c_2 and the range of change of ν [11]:

$$c_1 = 2.2797(\gamma+1)h^2c, \quad c_2 = 0, \quad |\nu| \leq 80.408779...^0. \tag{2.31}$$

It means that the unsymmetric term in the singular solution $\tilde{y}$ in (2.27) must be canceled, unlike the concept suggested by Cole and Cook [4]. Equations (2.24), (2.27), (2.30), and (2.31) show that in the leading term, the outer transonic small-disturbance flow field around the parabolic nose is symmetric about the x-axis (see also [8]).

With the dominant term in ϕ_1 being symmetric, it is expected that unsymmetric (circulation) effects may be represented by the second-order term in (2.23). The substitution of (2.23) back into (2.21) results in a homogeneous linear equation for the solution of the function f_p:

$$\left[f_\xi - \frac{36}{49}\xi^2\right] f_{p\xi\xi} + \left[f_{\xi\xi} + \frac{6}{7}\left(2a - \frac{13}{7}\right)\xi\right] f_{p\xi} - a(a-1)f_p = 0. \tag{2.32}$$

The solution of (2.32) shows [11] that the function f_p is an eigenfunction of the problem with an exponent,

$$a = \frac{4.2218878}{7}, \tag{2.33}$$

that satisfies $[\tilde{y}^a f_p(\xi)]_{\tilde{y}}(x, 0\pm) = 0$ as $x \to 0$ [11] and where

(2.34)

$$f_p = d_2(\sin\nu)^{0.3969} F\left(1.2036, -0.0370, \frac{3}{2}; \sin^2\nu\right) F^{-0.6031}\left(\frac{5}{3}, \frac{1}{2}, -\frac{3}{2}; \sin^2\nu\right).$$

Analysis of the second-order term shows that it represents a circulation flow around the leading edge of the airfoil, where d_2 is the circulation parameter that can be determined only from the complete numerical solution of the transonic small-disturbance problem (2.21).

Equations (2.23), (2.24), and (2.33) result in the approximation for (ϕ_1) in the nose region (as $x \to 0$):

$$(\gamma+1)\phi_1 \sim \tilde{y}^{4/7} f(\xi) + \tilde{y}^{4.2219/7} f_p(\xi) + \dots, \quad \xi = \frac{x}{\tilde{y}^{6/7}}. \tag{2.35}$$

Equation (2.35) shows that the transonic small-disturbance flow around the airfoil's leading edge is composed of a leading term that represents a basic symmetric flow due to the parabolic nose and a second-order term that represents an unsymmetric flow around the nose due to circulation. Although the two terms have different powers, it should be noted that these powers are relatively close. This specific structure of the outer transonic flow around the parabolic nose is different than the structure of the outer subsonic flow around the nose, where both symmetric and circulation effects appear in the same leading term (2.9).

The solution of the second-order problem (2.22) also results in a leading similarity term that represents a symmetric flow around the leading edge as $x \to 0$:

$$(\gamma+1)\phi_2 \sim \frac{1}{\gamma+1}\tilde{y}^{2/7} f_1(\xi) + \dots, \tag{2.36}$$

where the function $f_1(\xi)$ is also described by a parametric representation in terms of the hodograph-similarity variable ν (see [11], Eq. (68)).

As the leading edge of the airfoil is approached (as $x \to 0$), (2.23) to (2.36) show that the outer expansion of the transonic flow is described by

$$\begin{aligned} \Phi \sim U \Bigg\{ x &+ \frac{\delta^{2/3}}{\gamma+1}\left[\tilde{y}^{4/7} f(\xi) + \tilde{y}^{4.2219/7} f_p(\xi) + O(\tilde{y}^{4.4438/7})\right] \\ &+ \frac{\delta^{4/3}}{(\gamma+1)^2}\left[\tilde{y}^{2/7} f_1(\xi) + O(\tilde{y}^{4.2219/7})\right] + O(\epsilon(\delta)) \Bigg\}, \quad \xi = \frac{x}{\tilde{y}^{6/7}}. \end{aligned} \tag{2.37}$$

Equation (2.37) shows that in the leading-edge region as both $(x, \tilde{y}) \to 0$, the velocities in the x and $\tilde{y}$ directions become singular, specifically on the airfoil surface (as $\xi \to \infty$). This is known as the "transonic nose-singularity." As in the subsonic problem, there is also a misordering in the approximation (2.37) in the magnitude of the disturbances for every ξ when both x and y

are smaller than $\delta^2 h^2 c$. Therefore, a rescaling in the radial direction only is needed, $x^* = x/\delta^2, y^* = y/\delta^2$, in order to account for the local flow around the airfoil nose, where a stagnation point exists.

An inner limit process defines an expansion valid near the nose where $(\delta \to 0, M_\infty \to 1)$ with $1/M_\infty = 1 + K\delta^{2/3} + \dots$ and $(x^*, y^*; K, A)$ fixed and $x^* = x/\delta^2, y^* = y/\delta^2$. The form of the expansion is

$$\Phi(x, y; M_\infty, A, \delta) = U\delta^2 \phi_0(x^*, y^*)\,. \tag{2.38}$$

Because of the uniform scaling, ϕ_0 satisfies the full potential-flow equation (2.3) in the (x^*, y^*) plane; in the inner region, ϕ_0 describes a uniform flow with a sonic Mach number $(M_\infty = 1)$ about a parabola surface where the tangency boundary condition has to be satisfied (Fig. 2.7):

$$\phi_{0y^*} = \pm\frac{h\sqrt{c}}{\sqrt{x^*}}\phi_{0x^*} \quad \text{on} \quad y^* = \pm 2h\sqrt{cx^*}\,, \quad 0 \le x^* < \infty\,. \tag{2.39}$$

The expansion of the solution to the inner problem shows that as $|x^*| \to \infty$ or $|y^*| \to \infty$ [11],

$$\Phi \sim U\delta^2 \left\{ x^* + \frac{y^{*4/7}}{\gamma+1} f(\xi^*) + \frac{y^{*2/7}}{(\gamma+1)^2} f_1(\xi^*) + \dots \right\}, \tag{2.40}$$

where $\xi^* = x^*/y^{*6/7}$ is the inner similarity variable, and f, f_1 are the same functions as in the outer expansion (2.37).

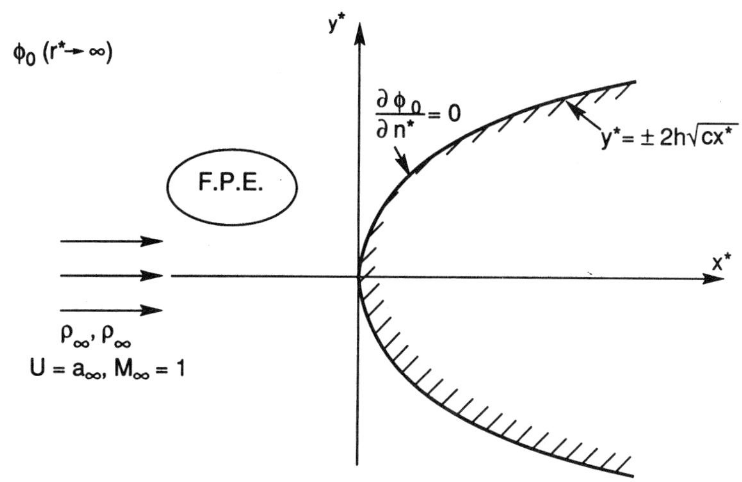

FIG. 2.7. *Parabolic nose problem (inner region).*

The matching between the near field of the outer expansion and the far field of the inner expansion is carried out with the help of an intermediate limit $(\delta \to 0, M_\infty \to 1)$ with $x_\eta = x/\eta(\delta), y_\eta = y/\eta(\delta)$ fixed. The region $\eta(\delta)$ is chosen such that $\delta^2 << \eta(\delta) << 1$, and as $\delta \to 0, \eta(\delta)/\delta^2 \to \infty$. Then $x = \eta(\delta)x_\eta \to 0$ and $\tilde{y} = \delta^{1/3}\eta(\delta)y_\eta \to 0$, whereas $|x^*| = \eta(\delta)/\delta^2|x_\eta| \to \infty$ and $|y^*| = \eta(\delta)/\delta^2|y_\eta| \to \infty$. Also, $\xi = x/\tilde{y}^{6/7} = (\eta/\delta^2)^{1/7}\xi_\eta$ and $\xi^* = x^*/y^{*6/7} = (\eta/\delta^2)^{1/7}\xi_\eta$ where $\xi_\eta = x_\eta/y_\eta^{6/7}$, so that in the intermediate region $\eta(\delta)$ as $\delta \to 0 : \xi = \xi^*$. Equations (2.37) and (2.40) show that the expansions read the same to a certain order when expressed in the (x_η, y_η) coordinates.

The matching process results in a well-defined boundary-value problem in the inner region for the solution of a compressible sonic flow around an infinite parabola surface $y^* = \pm 2h\sqrt{cx^*}$ at zero angle of attack. Since both the far-field approximation (2.40) and the boundary condition (2.39) are symmetric about the x-axis, $\phi_0(x^*, y^*)$ is also a symmetric function. The inner problem is solved numerically by using a tranformation to parabolic coordinates and a finite-differences scheme with a point over-relaxation algorithm. The numerical calculations provide the pressure distribution C_p^* along the parabolic nose (Fig. 2.8).

A uniformly valid solution for the potential Φ can be constructed from the outer transonic small-disturbance theory for ϕ_1 and the parabola inner solution for ϕ_0, by adding the two together and subtracting the common part in the intermediate region where the two solutions match. The composite solution is given in the limit $(\delta \to 0, M_\infty \to 1)$ with (K, K_1, A) fixed by

$$\begin{aligned}\Phi(x, y; M_\infty, A, \delta) \sim U\{x + \delta^{2/3}\phi_1(x, \tilde{y}; K, A) + \dots \\ + \delta^2\phi_0(x^*, y^*) + \dots - \Phi_{\text{c.p.}}(x_\eta, y_\eta; \delta, \eta(\delta))\},\end{aligned} \tag{2.41}$$

where

$$\Phi_{\text{c.p.}}(x_\eta, y_\eta; \delta, \eta(\delta)) = \eta x_\eta + \frac{1}{\gamma+1}\delta^{6/7}\eta^{4/7}y_\eta^{4/7} f\left(\left(\frac{\eta}{\delta^2}\right)^{1/7} \xi_\eta\right). \tag{2.42}$$

Using the composite solution and the isentropic relations, a uniformly valid pressure distribution over the entire airfoil is obtained. In the limit $(\delta \to 0, M_\infty \to 1)$ with (K, K_1, A) fixed

$$C_p(x; \delta, K, A) = C_p^*\left(\frac{x}{R_c}\right) + \frac{\rho^*}{\rho_\infty}\phi_{0x^*}\left[C_{p_{u,\ell}}(x; \delta, K, A) - C_{p\text{c.p.}}(x; \delta)\right]. \tag{2.43}$$

Here $C_p^*(x/R_c)$, ρ^*/ρ_∞, and ϕ_{0x^*} are calculated by the numerical solution of the sonic parabola problem, and $C_{p_{u,\ell}}(x; \delta, K, A)$ is the outer pressure coefficient on the upper and lower surfaces of the airfoil (that is calculated by the transonic small-disturbance theory). $C_{p\text{c.p.}}$ is the common part of the pressure coefficient

$$C_{p\text{c.p.}} = \frac{0.635776}{(\gamma+1)^{1/3}}\left[\frac{R_c}{x}\right]^{1/3}. \tag{2.44}$$

It coincides with the leading-edge singularity of the transonic small-disturbance pressure distribution $(C_{p_{u,\ell}})$ due to the leading effect of the parabolic nose of the airfoil [11].

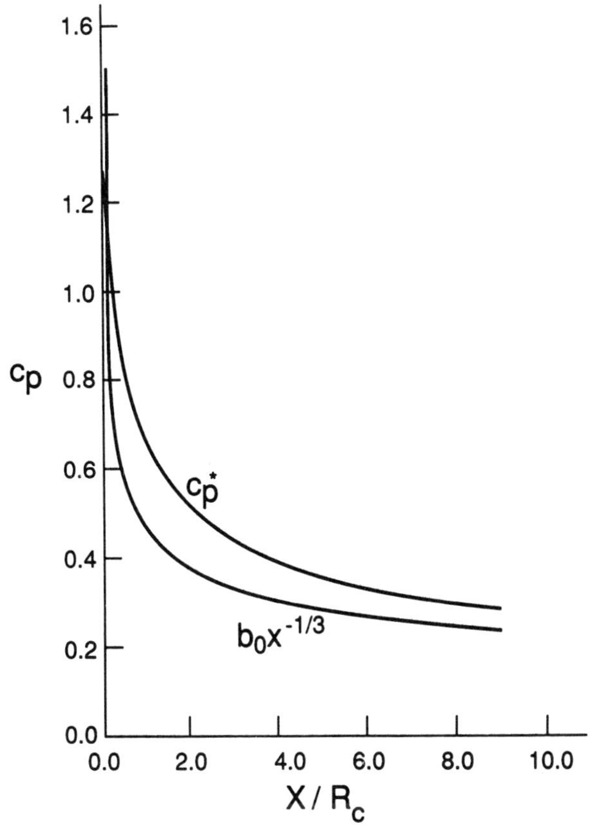

FIG. 2.8. *The pressure distribution along the parabola* (c_p^*) *and a comparison with the leading term of the pressure distribution from the transonic small-disturbance theory.*

Equation (2.43) behaves the same as (2.18) in the subsonic problem. As the leading edge of the airfoil is approached ($x \to 0$ or $0 < x < \delta^2 h^2 c$) the common pressure coefficient ($C_{\text{pc.p.}}$) cancels the ($x^{-1/3}$) singularity of the outer pressure coefficient both on the upper and lower surfaces of the airfoil. Also, in this region ϕ_{0x^*} is small and tends to zero near the stagnation point. Therefore, the dominant term in the leading-edge region is the parabola pressure coefficient C_p^*. The pressure distribution on the upper and lower surfaces of the airfoil is symmetric near the edge point, and to the orders discussed, the stagnation point is located at the leading edge, for every transonic Mach number of the oncoming flow and shape and angle of attack of the airfoil. As x is increased beyond the leading-edge region ($x > \delta h^2 c$), the density ρ^*/ρ_∞ and the velocity component ϕ_{0x^*} tend to one and the common pressure coefficient ($C_{\text{pc.p.}}$) cancels the parabola pressure coefficient (C_p^*). Therefore, the dominant term outside the leading-edge region is the outer pressure coefficient calculated by the transonic small-disturbance theory. In the intermediate re-

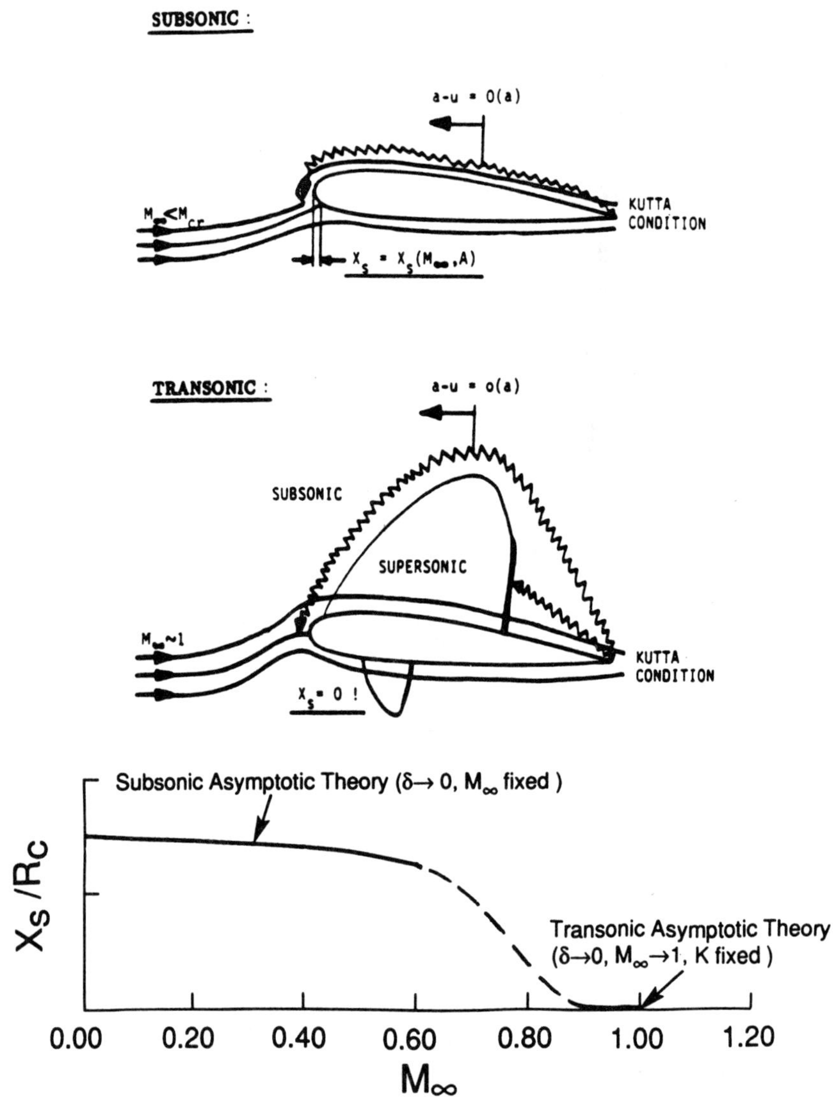

FIG. 2.9. *A suggested physical model.*

gion ($\delta^2 h^2 c < x < \delta h^2 c$) the pressure coefficient changes uniformly from (C_p^*) to ($C_{p_{u,\ell}}$) and asymmetric deviations from the symmetric inner solution increase and become significant only when the distance (x) from the leading edge of the airfoil increases beyond the inner region.

2.5. Discussion and a Physical Model

The comparison between the analyses of the subsonic and transonic flows indicates that the differences in the location of the stagnation point and in the change of the pressure distributions near the nose are due to basically different effects that govern the flows around the parabolic nose of the airfoil in each case. In the subsonic problem, the flow around the nose is determined mainly by the total circulation around the airfoil. On the other hand, in the transonic problem, the flow around the leading edge is governed mainly by the local shape of the parabolic nose and the circulation around the airfoil appears only as a second-order influence in the flow near the nose.

Based on these arguments, a possible physical mechanism is suggested to explain the differences between a subsonic and a transonic flow in the leading-edge region (Fig. 2.9). In the subsonic problem, the transonic flow around the airfoil is determined by the Kutta condition at the trailing edge of the airfoil which fixes the total circulation around the airfoil. The information about the Kutta condition travels upstream along the airfoil surfaces with a speed $(a-u)$ that has the order of the speed of sound (a_∞), reaches the leading edge of the airfoil, and strongly affects the flow around the nose and the location of the stagnation point. However, in the transonic problem where supersonic zones and shock waves exist in the flow around the airfoil, the information about the Kutta condition cannot penetrate the shock waves and the supersonic zones. Therefore, in order to reach the leading edge of the airfoil it has to bypass the supersonic zones and so has a longer way to go than in the subsonic problem. Also, it travels upstream with a speed $(a - u)$ that is much smaller than the speed of sound (a_∞). These two effects together prevent the information about the Kutta condition from reaching the leading-edge region as a first-order influence. Therefore, to the leading order, the transonic flow around the nose is locally governed by the shape of the parabolic nose. Since it is locally a symmetric shape, the flow around the nose is also locally symmetric and to the orders discussed, the stagnation point is located at the leading edge (the most forward point) of the airfoil for every transonic Mach number $(M_\infty \sim 1)$ and shape and small angle of attack of the airfoil.

Also shown in Fig. 2.9 is a schematic description of the shift of the stagnation point (x_s) along the parabolic nose of the airfoil as the Mach number (M_∞) is increased from zero to one for a fixed total circulation around the airfoil. As (M_∞) is increased, (x_s) tends toward the leading edge of the airfoil and is located there when $(M_\infty \sim 1)$. In the intermediate region where $M_\infty \sim M_{cr}$ the change of (x_s) with (M_∞) was not yet analyzed, but it is expected to have a uniform change as (M_∞) is changed from the subsonic to the transonic regime as has been recently calculated by Rusak and Webster [14]. The analysis of this problem seems to be as complicated as the above analyses of the subsonic and the transonic problems.

References

[1] I. H. Abbott and A. E. von Doenhoff, *Theory of Wing Sections*, Dover Publications Inc., New York, 1959, pp. 111–118.

[2] G. S. Baruzzi, W. G. Habashi, and M. M. Hafez, *Finite element solutions of the Euler equations for transonic external flows*, AIAA J., 29(11) (1991), pp. 1886–1893.

[3] H. Bateman, *Higher Transcendental Functions*, Vol. I, McGraw-Hill, New York, 1953.

[4] J. D. Cole and L. P. Cook, *Transonic Aerodynamics*, North-Holland, Amsterdam, 1986.

[5] I. Imai, *Application of the M^2-expansion method to the subsonic flow of a compressible fluid past a parabolic cylinder*, Proc. 1st Japan Congress for Appl. Mechanics, 1951, pp. 349–352.

[6] A. Jameson, *Transonic Flow Calculation*, Princeton University MAE Report #1651, 1985.

[7] R. T. Jones, *Leading-edge singularities in thin-airfoil theory*, J. Aeronaut. Sci., 17 (1950), pp. 307–310.

[8] B. L. Keyfitz, R. E. Melnik, and B. Grossman, *An analysis of the leading-edge singularity in transonic small disturbance theory*, Quart. J. Mech. Appl. Math., XXXI, Pt. 2 (1978), pp. 137–155.

[9] E. A. Müller and K. Matachat, *Ägbkucgjeutskösungen der Transsonishen gleichumgen bai der Anstroöm – Machzahl 1*, Proc. Eleventh International Congress of Applied Mechanics, Munich, 1964, pp. 1064–1068.

[10] Z. Rusak, *Subsonic Flow Around a Thin Airfoil With a Parabolic Nose*, Department of Mathematical Sciences Report No. 186, Rensselaer Polytechnic Institute, Troy, New York, 1990. European J. Appl. Math., in press.

[11] ——, *Transonic Flow Around the Leading Edge of a Thin Airfoil with a Parabolic Nose*, Department of Mathematical Sciences Report No. 188, Rensselaer Polytechnic Institute, Troy, New York, 1990. European J. Appl. Math., in press, 1993.

[12] ——, *Similarity Solutions of the Sonic Small-Disturbance Equation*, Department of Mathematical Sciences Report No. 189, Rensselaer Polytechnic Institute, Troy, New York, 1991. Submitted to SIAM J. Appl. Math.

[13] ——, *The Sonic Flow about Two-Dimensional Slender Semi-Infinite Bodies*, Department of Mathematical Sciences Report No. 196, Rensselaer Polytechnic Institute, Troy, New York, 1991.

[14] Z. Rusak and B. E. Webster, *The subsonic and transonic flow around the leading edge of a thin airfoil with a parabolic nose*, AIAA paper 92–2649, AIAA 10th Applied Aerodynamics Conference, Palo Alto, 1992.

[15] M. D. Van Dyke, *Subsonic Edges in Thin-Wing and Slender-Body Theory*, NACA TN3343, Nov. 1954.

[16] ——, *Perturbation Methods in Fluid Mechanics*, Parabolic Press, Stanford, CA, 1975, pp. 54–56, 59–66.

Chapter 3

An Asymptotic Approach to Separation and Stability Problems of a Transonic Boundary Layer

Oleg S. Ryzhov

3.1. The State of the Art

The asymptotic approach under discussion is based on the notion of an interacting boundary layer with the self-induced pressure gradient not known in advance. More specifically, an extension of the triple-deck theory invented by Neiland, Stewartson, and Messiter will be considered and employed to predict stability properties of the Blasius boundary layer on a thermally insulated flat plate in the transonic range of external velocities. However, before we set about to solve this problem it is instrumental to briefly outline the fundamental features of the triple deck with regard to subsonic and supersonic boundary layers. In these cases the unique small parameter $\epsilon = R^{-1/8}$—based on the Reynolds number $R = \rho_\infty^* U_\infty^* L^*/\lambda_\infty^* \to \infty$ where $\rho_\infty^*, \lambda_\infty^*$, and U_∞^* are the density, viscosity, and velocity of an oncoming uniform stream, and L^* is the characteristic length of the body, respectively—comes into play and, importantly, it can be ruled out from the final formulation of the boundary-value problem through a suitable definition of physical quantities. The sketch of the triple deck is shown in Fig. 3.1 along with the corresponding scalings of different sublayers in longitudinal and transverse directions in terms of ϵ. The outer region I is occupied by a potential flow; disturbances in the main deck II are inviscid in nature in spite of the essentially viscous pattern of the initial Blasius boundary layer; vorticity waves in the thin wall sublayer III are governed by the balance between the inertial forces, pressure, and viscous stresses. If the difference $\hat{\delta} = M_\infty^2 - 1 = O(1)$ is properly included in scalings for both independent variables and unknown functions, the final formulation of the boundary value problem also becomes free from dependence on the Mach number M_∞.

To clarify these statements let us go into some details of subsonic and supersonic flows. In the viscous wall sublayer an appropriate definition of

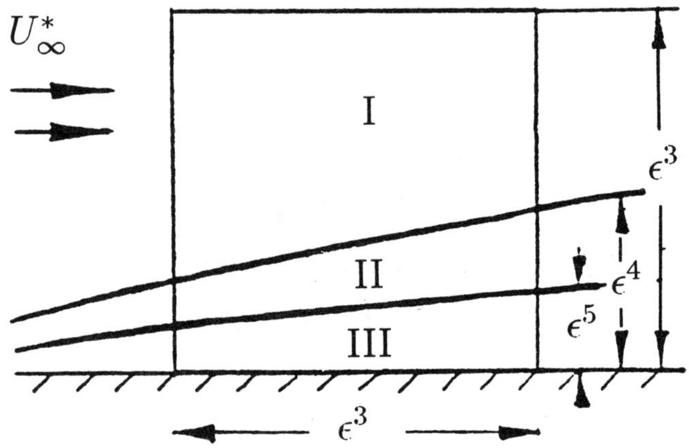

FIG. 3.1. *General sketch of the local triple-deck structure of an interacting boundary layer.*

the time t^*, longitudinal x^*, and transverse y^* cartesian coordinates reads as follows

$$t^* = \frac{L^*}{U_\infty^*}\epsilon^2 \mid \hat{\delta} \mid^{-1/4} \beta^{-1} t, \tag{3.1a}$$

$$x^* = L^*(1+\epsilon^3 \mid \hat{\delta} \mid^{-3/8} x), \qquad y^* = L^*\epsilon^5 \mid \hat{\delta} \mid^{-1/8} y, \tag{3.1b, c}$$

whereas the corresponding components u^* and v^* of the perturbed velocity vector along with the difference $p^* - p_\infty^*$ between the local p^* and the equilibrium p_∞^* pressures are normalized by

(3.2a, b)

$$u^* = U_\infty^*[\epsilon \mid \hat{\delta} \mid^{-1/8} u(t,x,y) + ...], \qquad v^* = U_\infty^*[\epsilon^3 \mid \hat{\delta} \mid^{1/8} v(t,x,y) + ...],$$

$$p^* - p_\infty^* = \rho_\infty^* U_\infty^{*2}[\epsilon^2 \mid \hat{\delta} \mid^{-1/4} p(t,x,y) + ...], \tag{3.2c}$$

the constant $\beta = O(1)$ being measured in terms of the nondimensional frequency. In the asymptotic limit $R \to \infty$ we then have for the both regimes the Prandtl equations

$$\frac{\partial u}{\partial x} + \frac{\partial v}{\partial y} = 0, \qquad \frac{\partial p}{\partial y} = 0, \tag{3.3a, b}$$

$$\left(\frac{T_w^*}{T_\infty^*}\right)^{-1}\left(\beta\frac{\partial u}{\partial t} + u\frac{\partial u}{\partial x} + v\frac{\partial u}{\partial y}\right) = -\frac{\partial p}{\partial x} + C\frac{T_w^*}{T_\infty^*}\frac{\partial^2 u}{\partial y^2} \tag{3.3c}$$

describing the motion of an effectively incompressible fluid [8], [18], [19], [6]. Here T_w^* and T_∞^* designate the wall and free stream temperatures, respectively;

C is the constant entering the Chapman law $\lambda_w^*/\lambda_\infty^* = CT_w^*/T_\infty^*$ for the viscosity distribution over the body surface. Additional parameters of order 1 appearing in the last equation could be excluded through simple rescaling of the independent variables and unknown functions but it does not matter at this stage of analysis.

At the plate surface we have, as usual, the no-slip conditions

$$u = v = 0 \qquad \text{at } y = 0, \tag{3.4}$$

but when approaching the upper reaches of the wall sublayer a new unknown function, the normalized displacement thickness $-A$, enters into the limit condition

$$u - \tau_w C^{-1/2}(T_w^*/T_\infty^*)^{-1} y \to \tau_w C^{-1/2}(T_w^*/T_\infty^*)^{-1} A \qquad \text{as } y \to \infty, \tag{3.5}$$

$\tau_w = 0.3321$ being a nondimensional wall friction in the Blasius boundary layer.

To make the boundary value problem complete we need a relationship to connect the excess pressure with the displacement thickness. Obviously, the precise form of this relationship should depend strongly on the flow pattern in the upper deck where it is determined by the small disturbance theory. Thus, the Laplace equation can be applied for deriving the velocity component distributions if $M_\infty < 1$ and the wave equation in space variables gives a basis if $M_\infty > 1$. As a result the relationship sought for a subsonic boundary layer is expressed in terms of an integral [18], [6]

$$p = \frac{1}{\pi} \int_{-\infty}^{\infty} \frac{\partial A(t,\xi)/\partial \xi}{x - \xi} d\xi \tag{3.6}$$

whereas a much simpler formula [8], [19]

$$p = -\frac{\partial A}{\partial x} \tag{3.7}$$

holds true when a boundary layer adjoins a uniform supersonic stream at infinity.

Three comments are relevant at this stage. First the disturbance propagation in the potential part of the flow is always quasi-steady and, accordingly, both relations (3.6) and (3.7) are independent of time. Second, the boundary value problem as posed forms a basis for a nonlinear study in general, the linear approach being its simplest version. Third, in the framework of the triple-deck theory it is impossible to calculate the critical value of the Reynolds numbers. On the contrary, the same reasoning can be used for qualitative study of the major features of oscillations with the amplitude growing in time and space. So, the asymptotic analysis under discussion is aimed at revealing general rules defining the process in kind rather than in degree.

Let us proceed now to developing a linear approach. As mentioned above, all four nondimensional parameters β, C, τ_w, and T_w^*/T_∞^* of order 1 can be eliminated from the time-scaling (3.1a), equation (3.3c), and boundary condition (3.5) as a result of an additional affine transformation. The simplest way is to put $\beta = 1$ and then introduce rescaled independent variables

(3.8a) $$t = C^{1/4}\tau_w^{-3/2}(T_w^*/T_\infty^*)t',$$

(3.8b, c) $$x = C^{3/8}\tau_w^{-5/4}(T_w^*/T_\infty^*)^{3/2}x', \qquad y = C^{5/8}\tau_w^{-3/4}(T_w^*/T_\infty^*)^{3/2}y',$$

along with modified unknown functions

(3.9a, b) $$u = C^{1/8}\tau_w^{1/4}(T_w^*/T_\infty^*)^{1/2}u', \qquad v = C^{3/8}\tau_w^{3/4}(T_w^*/T_\infty^*)^{1/2}v',$$

(3.9c, d) $$p = C^{1/4}\tau_w^{1/2}p', \qquad A = C^{5/8}\tau_w^{-3/4}(T_w^*/T_\infty^*)^{3/2}A'.$$

Keeping the former notation for new quantities we linearize the problem around the uniform shear

(3.10) $$u = y, \qquad v = p = A = 0,$$

which gives the undisturbed flow in the viscous wall sublayer, and separate the t- and x-dependence of eigenoscillations through

(3.11) $$(u - y, v, p, A) = a[u_0(y), v_0(y), p_0, A_0]\exp[i(\omega t + kx)],$$

where $a \to 0$, ω, and k are the amplitude, frequency, and wavenumber, respectively. Then, upon substituting (3.11) into the Prandtl equations and all the boundary conditions, the following dispersion relation

(3.12a) $$\Phi(\Omega) = Q(k),$$

(3.12b, c) $$\Phi = \frac{dAi(\Omega)}{dz}\left[\int_\Omega^\infty Ai(z)dz\right]^{-1}, \qquad \Omega = i^{1/3}\omega k^{-2/3},$$

connecting the frequencies and wavenumbers arises, $Ai(z)$ being the Airy function. Whereas the left-hand side $\Phi(\Omega)$ is the same for both regimes under consideration, the function $Q(k)$ in the right-hand side, depending on the wavenumber k only, is given by

(3.13) $$Q = i^{1/3}k^{1/3} \mid k \mid$$

if the oncoming stream is subsonic and expressed through

(3.14) $$Q = -i^{4/3}k^{4/3}$$

in the case of the supersonic stream in the potential region. The complex variable z reads as $z = \Omega + i^{1/3}k^{1/3}y$; a cut in the k-plane should be drawn along the positive imaginary semi-axis to fix a regular branch of the function $k^{1/3}$, i.e., $-3\pi/2 \leq \arg k \leq \pi/2$. The properties of the dispersion relation are well studied and documented (see, for example, [15], [20], [14].

The distinction between the two expressions (3.13) and (3.14) for Q is crucial. For a subsonic boundary layer, among the infinite set of the dispersion relation roots $\Omega_j = \Omega_j(k)$ providing the dispersion curves $\omega_j = \omega_j(k)$ the first one with $j = 1$ can lead to unstable oscillations. The trace of the first root ω_1 in the complex ω-plane with k running through real positive values is shown in Fig. 3.2. The part of the curve located in the upper half-plane represents stable pulsations, while the part of the same curve from the lower half-plane gives rise to unstable disturbances. The critical frequency $\omega = \omega'_* = -2.298$ and wavenumber $k = k'_* = 1.0005$ of the neutral oscillations being fixed by $\text{Re}(\omega_1(k)) = 0$. It is worth stressing that according to Smith [16] and Zhuk and Ryzhov [21] all of these types of disturbances are asymptotic forms of Tollmien-Schlichting waves in the limit as $R \to \infty$ and derivable from the analysis of the Orr-Sommerfeld equation.

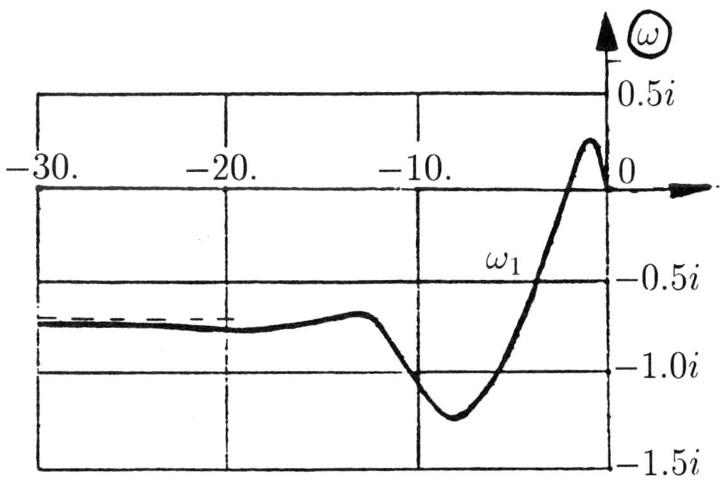

FIG. 3.2. *The trace of the first root* $\omega = \omega_1(k)$ *in the complex* ω*-plane with* k *running through real positive values for a subsonic boundary layer. The right-hand side of the dispersion relation* (3.12a–c) *is given by* (3.13).

On the contrary, for a supersonic boundary layer, none of the dispersion relation roots $\Omega_j = \Omega_j(k)$ is associated with unstable disturbances. The traces of first three roots $\omega_j = \omega_j(k)$ in this case are plotted in Fig. 3.3. They pass through the second quadrant of the complex ω-plane if k takes on real positive values. We may conclude hence that a supersonic boundary layer is stable in the framework of the triple-deck theory. This prediction obviously contradicts both measurements and calculations.

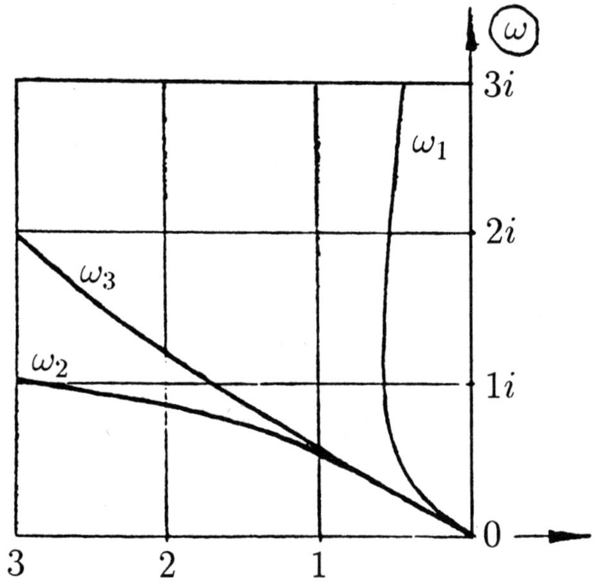

FIG. 3.3. *The traces of the first three roots* $\omega = \omega_j(k), j = 1, 2, 3$ *in the complex* ω*-plane with* k *running through real positive values for a supersonic boundary layer. The right-hand side of the dispersion relation* (3.12a–c) *is given by* (3.14).

3.2. General Properties of Two Transonic Regimes

We are now in a position where the same asymptotic theory predicts the proper limit of the eigenoscillation behavior if $M_\infty < 1$ and is at variance with all the data and findings available for $M_\infty > 1$. The problem is to extend the theory of the subsonic boundary layer instability to the transonic and slightly supersonic range of external velocities rather than to fill a gap between two different regimes described to the same degree of accuracy. This objective is obviously distinct from spanning with a bridge two different applicability ranges of the triple deck.

To provide a basis for understanding unsteady processes initiated by disturbance propagation in transonic boundary layer, we need to reconsider the outer potential region of the velocity field. On the assumption that $\mid M_\infty^2 - 1 \mid \ll 1$, the second small parameter

$$\delta' = (M_\infty^2 - 1)/K'_\infty \tag{3.15}$$

appears here, K'_∞ being a positive constant of order 1. The substitution δ' for $\hat{\delta}$ is aimed at providing a suitable entity for measuring the relative magnitude of ϵ and $\mid M_\infty^2 - 1 \mid$; (3.15) is meant to be introduced in all normalized variables employed above. Let us analyze now in more detail the consequences of conventional scalings (3.1a,b) for the time t^* and longitudinal coordinate x^*, respectively, as well as the definitions of the normal coordinate

$$y^* = L^* \epsilon^3 \mid \delta' \mid^{-7/8} y_1 \tag{3.16}$$

and the potential function

$$\varphi^* = U^*_\infty L^*[x + \epsilon^5 \mid \delta' \mid^{-5/8} \varphi_1(t, x, y_1) + ...] \tag{3.17}$$

in the external part of the velocity field. In what follows the product $\epsilon \mid \delta' \mid^{-1/8}$ is supposed to be much less than 1. Then an equation for φ_1 to appear first in [5] reads

$$\begin{aligned} &2\beta\frac{\partial^2\varphi_1}{\partial t\partial x} + \epsilon^{-1} \mid \delta' \mid^{1/8} \left(\delta' K'_\infty + 2\epsilon^2 \mid \delta' \mid^{-1/4} \frac{\partial\varphi_1}{\partial x}\right) \frac{\partial^2\varphi_1}{\partial x^2} \\ &- \epsilon^{-1} \mid \delta' \mid^{9/8} \frac{\partial^2\varphi_1}{\partial y_1^2} = 0. \end{aligned} \tag{3.18}$$

An attempt to retain all the terms here leads to estimations $\mid \delta' \mid = O(\epsilon^{8/5})$ and $\beta = O(\epsilon^{4/5})$. The full equation (3.18) is relevant if we deal with an essentially nonlinear process in the outer region such as slow motion or vibrations of a shock. Taking the first alternative the governing equation can be rewritten as

$$2\beta\epsilon \mid \delta' \mid^{-9/8} \frac{\partial^2\varphi_1}{\partial t\partial x} + \left(\pm K'_\infty + 2\epsilon^2 \mid \delta' \mid^{-5/4} \frac{\partial\varphi_1}{\partial x}\right) \frac{\partial^2\varphi_1}{\partial x^2} - \frac{\partial^2\varphi_1}{\partial y_1^2} = 0, \tag{3.19a}$$

$$K'_\infty = O[|M^2_\infty - 1|/\epsilon^{8/5}], \tag{3.19b}$$

with K'_∞ for positive δ' and $-K'_\infty$ if δ' is supposed to be negative. However, with $\beta = O(\epsilon^{4/5}) \to 0$ the time-derivative $\partial u/\partial t$ drops out of the Prandtl equation (3.3c) for the viscous wall sublayer where the velocity (and vorticity) fields turn out to be quasi-steady and adjust instantly to the propagation of disturbances driven by pressure variations in the potential flow. This approach is applicable also in investigating phenomena in which viscosity plays a decisive part, separation being the most important among them.

On the other hand, if the time-derivative $\partial u/\partial t$ does not enter the Prandtl equations their solutions do not include the eigenoscillations of the boundary layer. For elucidating the stability properties we should choose the second alternative given by $\beta = O(1)$. Hence it follows that $\mid \delta' \mid = O(\epsilon^{8/9})$. The nonlinear term $\partial\varphi_1/\partial x \cdot \partial^2\varphi_1/\partial x^2$ becomes negligible under such conditions as compared with the linear ones and can be disregarded. Then the governing equation (3.18) reduces to

$$2\beta\epsilon \mid \delta' \mid^{-9/8} \frac{\partial^2\varphi_1}{\partial t\partial x} \pm K'_\infty \frac{\partial^2\varphi_1}{\partial x^2} - \frac{\partial^2\varphi_1}{\partial y_1^2} = 0, \tag{3.20a}$$

$$K'_\infty = O[|M^2_\infty - 1|/\epsilon^{8/9}], \tag{3.20b}$$

where again K'_∞ is to be considered for positive δ' and $-K'_\infty$ appears in the case of negative δ'. Obviously, the constant K'_∞ is nothing more than a similarity parameter taking on different forms (3.19b) and (3.20b) for two different transonic regimes. Precise definitions of the similarity parameter for the both cases that incorporate also the dependence on C, τ_w, and T_w^*/T_∞^* will be given in the next sections.

To match solutions in the outer region and in the viscous wall sublayer we need to consider most of the boundary layer. The latter plays a passive role and serves just for striking a balance between the pressure variations in the potential part of the flow and the corresponding response in the displacement thickness provided by the viscous wall sublayer in order to maintain the necessary slope of streamlines. As a result of matching, the explicit relations (3.6) and (3.7) to directly connect the excess pressure p with the displacement thickness $-A$ have arisen in the above analysis for subsonic and supersonic boundary layers, respectively. For both types of transonic flows the matching conditions can be written as

$$\frac{\partial \varphi_1(t,x,0)}{\partial x} = -p(t,x), \qquad \frac{\partial \varphi_1(t,x,0)}{\partial y_1} = -\frac{\partial A(t,x)}{\partial x}, \tag{3.21a, b}$$

at the bottom, $y_1 = 0$, of the potential region. Concerning the limit condition for the upper reaches of the viscous wall sublayer, i.e., as $y \to \infty$, it remains unchanged and given by (3.5).

3.3. Separation

Thus, to evaluate properly in more detail the general concepts exposed with regard to the existence of two types of transonic flows, it is advantageous to use the group properties of the equations and boundary conditions for an additional renormalization of all physical quantities. If the first transonic regime is to be considered then the following affine transformation

$$t = 2^{-1/10} C^{1/10} \tau_w^{-9/5} (T_w^*/T_\infty^*)^{2/5} t', \tag{3.22a}$$

(3.22b, c)

$$x = 2^{-3/10} C^{3/10} \tau_w^{-7/5} (T_w^*/T_\infty^*)^{6/5} x', \qquad y = 2^{-1/10} C^{3/5} \tau_w^{-4/5} (T_w^*/T_\infty^*)^{7/5} y',$$

$$y_1 = 2^{-7/10} C^{1/5} \tau_w^{-8/5} (T_w^*/T_\infty^*)^{4/5} y_1', \tag{3.22d}$$

for independent variables and

(3.23a, b)

$$u = 2^{-1/10} C^{1/10} \tau_w^{1/5} (T_w^*/T_\infty^*)^{2/5} u', \qquad v = 2^{1/10} C^{2/5} \tau_w^{4/5} (T_w^*/T_\infty^*)^{3/5} v',$$

(3.23c, d)

$$p = 2^{-1/5} C^{1/5} \tau_w^{2/5} (T_w^*/T_\infty^*)^{-1/5} p', \qquad A = 2^{-1/10} C^{3/5} \tau_w^{-4/5} (T_w^*/T_\infty^*)^{7/5} A',$$

(3.23e) $$\varphi_1 = 2^{-1/2}C^{1/2}\tau_w^{-1}(T_w^*/T_\infty^*)\varphi_1',$$

for unknown functions along with $\beta = \epsilon^{4/5}$ and $|\ \delta'\ | = \epsilon^{8/5}$ taken for the sake of simplicity reduces the boundary-value problem to a canonical form. Combining the results of two preceding sections yields ([7], [1], [9])

(3.24a, b, c) $$\frac{\partial u}{\partial x} + \frac{\partial v}{\partial y} = 0, \qquad u\frac{\partial u}{\partial x} + v\frac{\partial u}{\partial y} = -\frac{\partial p}{\partial x} + \frac{\partial^2 u}{\partial y^2}, \qquad \frac{\partial p}{\partial y} = 0,$$

(3.24d) $$u - y \to A(t, x) \qquad \text{as } y \to \infty,$$

(3.24e) $$u = v = 0 \qquad \text{at } y = 0,$$

(3.24f) $$\frac{\partial^2 \varphi_1}{\partial t \partial x} + \left(K_\infty + \frac{\partial \varphi_1}{\partial x}\right)\frac{\partial^2 \varphi_1}{\partial x^2} - \frac{\partial^2 \varphi_1}{\partial y_1^2} = 0,$$

(3.24g, h) $$\frac{\partial \varphi_1}{\partial x} = -p(t, x)\,, \quad \frac{\partial \varphi_1}{\partial y_1} = -\frac{\partial A}{\partial x} \quad \text{at} \quad y_1 = 0,$$

with primes omitted in this notation. The only similarity parameter

(3.25) $$K_\infty = \frac{M_\infty^2 - 1}{\epsilon^{8/5}2^{4/5}C^{1/5}\tau_w^{2/5}(T_w^*/T_\infty^*)^{4/5}}$$

making (3.19b) more precise defines the nature of disturbances within the nonlinear potential velocity field and the steady vorticity distribution within the viscous wall sublayer. The equation (3.24f) is typical of transonic aerodynamics (see, for example, [2]).

The flow pattern around a small hump immersed into the thin wall sublayer on an otherwise flat plate serves as an example to illustrate the occurrence of separation. In this case simultaneous integration of (3.24a–c) and (3.24f) is subjected to the above boundary conditions (3.24d), (3.24g,h), and the no-slip conditions $u = v = 0$ on the obstacle surface $y = y_w(x)$ rather than at $y = 0$. Besides, the additional requirements $u \to 0$, $p \to 0$ should be met at the entry, $x \to -\infty$, into the viscous sublayer. The results of numerical calculations presented in Fig. 3.4a,b show the streamline shapes in the vicinity of the hump $y_w = h(1 + \cos x), h = 2, -\pi \le x \le \pi$ for two values of $K_\infty = \mp 3$ [10]. According to them the velocity field changes drastically when the subcritical regime is superseded by the supercritical one. In the former case, separation takes place first on the leeward side of the hump, whereas the latter case is marked by the development of a much smaller recirculation bubble on its windward side. It should be kept in mind that these structures are local and provoked actually by a small roughness on the plate surface. Obviously they

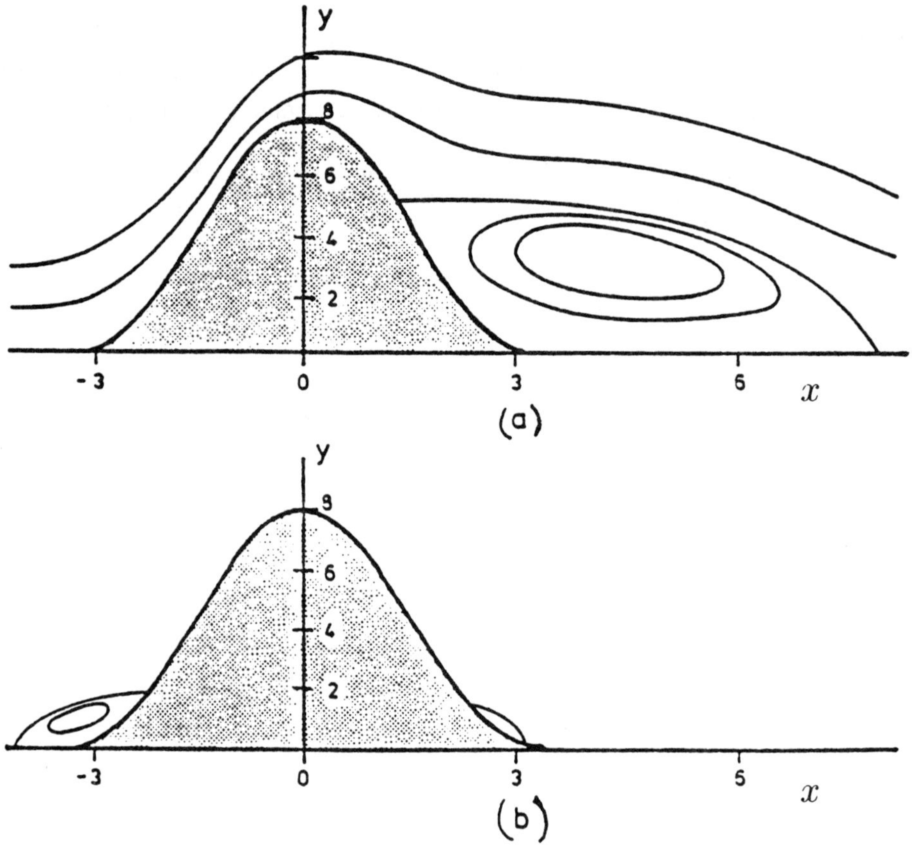

FIG. 3.4. *Transonic flow pattern around a hump* $y_\omega = h(1+\cos x), h = 2, -\pi \leq x \leq \pi$ *on an otherwise flat plate: (a)* $K_\infty = -3$*; (b)* $K_\infty = 3$.

are not directly connected to the global flow pattern around an airfoil in the oncoming transonic stream when the boundary layer separates, as a rule, under the influence of the pressure rise across a shock front terminating an extensive supersonic pocket. On the other hand, the local separation under discussion can promote the onset of a more massive one. Additional calculations are needed to clear up the essence of this phenomenon.

Let us inquire briefly into how the boundary value problem posed settles the issue of the Blasius boundary layer response to high frequency disturbances within the linear approximation. For this purpose the pressure perturbations in the potential field can be assumed to be of the form

(3.26a, b, c)

$$p = a\exp[i(\omega t + kx) - \lambda(\omega\, k)y_1], \qquad \lambda = D^{1/2}, \qquad D = -\omega k - K_\infty k^2,$$

consistent with both the linearized equation (3.24f) and boundary conditions (3.24g,h) where the function λ is a branch of $(-\omega k - K_\infty k^2)^{1/2}$ with positive real part. The excess pressure (3.26a–c) gives rise to the boundary layer oscillations expressed by (3.11). The dispersion relation to connect the frequency ω with the wavenumber k reads

(3.27a, b, c)
$$\Phi(0) = Q(\omega, k), \qquad \Phi(0) = -\frac{3^{7/6}\Gamma(2/3)}{2\pi}, \qquad Q(\omega, k) = \frac{(ik)^{1/3}k^2}{\lambda}.$$

With due account of constraints on regular branches of $k^{1/3}$ and λ this relation results in $-5\pi/7 \leq \arg k \leq -2\pi/7$. The latter implies that the pressure at the outer edge of the viscous wall sublayer cannot be specified, at an initial instant $t = 0$, as a harmonic function of x to allow tracing the ensuing evolution of oscillations in time. This is a distinctive feature of the disturbances at hand that makes them completely dissimilar to the Tollmien-Schlichting waves of the hydrodynamic stability theory [10]. On the other hand, the value $\arg k = -\pi/2$ to fix the upstream influence in steady supersonic flows with $K_\infty > 1$ [4] lies within the range of $\arg k$ indicated.

3.4. Stability

Thus, to evaluate properly the unsteady response of the flow as a whole to initial harmonic disturbances we need time-dependence to be introduced in the viscous wall sublayer. Hence the second transonic regime becomes operative, in which case the additional affine transformation

(3.28a)
$$t = 2^{-2/9}C^{2/9}\tau_w^{-14/9}(T_w^*/T_\infty^*)^{8/9}t',$$

(3.28b, c)
$$x = 2^{-3/9}C^{1/3}\tau_w^{-4/3}(T_w^*/T_\infty^*)^{4/3}x', \quad y = 2^{-1/9}C^{11/18}\tau_w^{-7/9}(T_w^*/T_\infty^*)^{13/9}y',$$

(3.28d)
$$y_1 = 2^{-7/9}C^{5/18}\tau_w^{-13/9}(T_w^*/T_\infty^*)^{10/9}y_1',$$

for independent variables and

(3.29a, b)
$$u = 2^{-1/9}C^{1/9}\tau_w^{2/9}(T_w^*/T_\infty^*)^{4/9}u', \quad v = 2^{1/9}C^{7/18}\tau_w^{7/9}(T_w^*/T_\infty^*)^{5/9}v',$$

(3.29c, d)
$$p = 2^{-2/9}C^{2/9}\tau_w^{4/9}(T_w^*/T_\infty^*)^{-1/9}p', \quad A = 2^{-1/9}C^{11/18}\tau_w^{-7/9}(T_w^*/T_\infty^*)^{13/9}A',$$

(3.29e)
$$\varphi_1 = 2^{-5/9}C^{5/9}\tau_w^{-8/9}(T_w^*/T_\infty^*)^{11/9}\varphi_1',$$

for unknown functions together with $\beta = 1$ and $\mid \delta' \mid = \epsilon^{8/9}$ reduces the boundary value problem to another canonical representation as distinguished from (3.24a–h). The final results have been summarized by Ryzhov and Savenkov [11] as follows

(3.30a, b)
$$\frac{\partial u}{\partial x} + \frac{\partial v}{\partial y} = 0, \qquad \frac{\partial p}{\partial y} = 0,$$

(3.30c) $$\frac{\partial u}{\partial t}+u\frac{\partial u}{\partial x}+v\frac{\partial u}{\partial y}=-\frac{\partial p}{\partial x}+\frac{\partial^2 u}{\partial y^2},$$

(3.30d) $$u-y\to A(t,x) \quad \text{as} \quad y\to\infty,$$

(3.30e) $$u=v=0 \quad \text{at} \quad y=0,$$

(3.30f) $$\frac{\partial^2\varphi_1}{\partial t\partial x}+K_\infty\frac{\partial^2\varphi_1}{\partial x^2}-\frac{\partial^2\varphi_1}{\partial y_1^2}=0,$$

(3.30g, h) $$\frac{\partial\varphi_1}{\partial x}=-p(t,x), \quad \frac{\partial\varphi_1}{\partial y_1}=-\frac{\partial A}{\partial x} \quad \text{at} \quad y_1=0,$$

with primes in notation omitted again for the sake of simplicity. The similarity parameter

(3.31) $$K_\infty=\frac{M_\infty^2-1}{\epsilon^{8/9}2^{8/9}C^{1/9}\tau_w^{2/9}(T_w^*/T_\infty^*)^{4/9}}$$

making (3.20b) more precise completely defines the disturbance pattern on a flat plate $y=0$. Since oscillations will be assumed further on to be harmonic in space, no additional constraints at the entry, $x\to-\infty$, into the viscous sublayer should be required.

Two comments are relevant with regard to the boundary value problem posed. The linearization of the external potential field is a crucial standpoint for transonic aerodynamics where the common trend is to retain the nonlinear description (see, for example, [2]). On the other hand, the essentially unsteady nature of disturbances propagating through the external region is a new key element introduced into the conventional version of the triple-deck theory that operates as a rule with time-independent relations connecting the excess pressure with the displacement thickness ([16], [21]). Owing to the different affine transformations employed for two kinds of transonic regimes, the similarity parameters (3.25) and (3.31) are based not only on different powers of the small parameter ϵ but include also different powers of the nondimensional constants C and τ_w as well as the ratio T_w^*/T_∞^*.

To carry out a test as to whether the new approach expressed in (3.30a–h) is capable of predicting the loss of stability in the transonic range of velocities, let us linearize the full problem which is still nonlinear owing to the Prandtl equation (3.30c). In addition to (3.11) separating t- and x-dependence of eigenoscillations in the viscous wall sublayer, a similar representation

(3.32) $$\varphi_1=a\varphi_0(y_1)\exp[i(\omega t+kx)]$$

for the potential function is to be considered simultaneously. As a result of a conventional procedure, where the matching conditions (3.30g,h) allow coupling both disturbance fields, the dispersion relation arises in the form

$$\Phi(\Omega) = Q(\omega, k), \tag{3.33}$$

which is composed of the left-hand side (3.12b,c) of the dispersion relation for subsonic and supersonic boundary layers and the right-hand side (3.27c), with λ as in (3.26b,c), of the dispersion relation for a transonic boundary layer in the first regime. This brings to light again the role of nonlinearity of the process in the upper irrotational sublayer that makes the wave propagation too fast to be accompanied by the unsteady response of vorticity pulsations in the lower viscous sublayer. Thus, a distinctive feature of both transonic regimes can be attributed to the frequency ω appearing in Q as the second argument.

It is necessary now to clear up the issue of whether the feature just mentioned is sufficient for achieving a smooth passage through $M_\infty = 1$ when we are approaching this threshold value of the Mach number from below. Since a supersonic boundary layer is, according to §3.1, stable against harmonic disturbances with arbitrary wavenumbers, the simplest way to make the statement convincing (but not rigorous, certainly) is to show the existence of a neutral eigenmode with the amplitude remaining constant in time even if the similarity parameter $K_\infty > 1$. Let ω_* and k_* be the frequency and wavenumber of this presumable mode in the transonic boundary layer under consideration. The analogous quantities $\omega'_* = -2.298$ and $k'_* = 1.0005$ have been evaluated in §3.1 for the subsonic boundary layer. In this notation the dispersion relation (3.33) reduces to

$$\frac{\omega_*}{k_*^{2/3}} = \frac{\omega'*}{k'^{\,2/3}_*}, \qquad \frac{k_*^{7/3}}{(-\omega_* k_* - K_\infty k_*^2)^{1/2}} = k'^{\,4/3}_*. \tag{3.34a, b}$$

The limit process $K_\infty \to -\infty$ yields the asymptotics

$$\omega_* \to \omega'_* \mid K_\infty \mid^{1/4} + \ldots, \qquad k_* \to k'_* \mid K_\infty \mid^{3/8} + \ldots \tag{3.35a, b}$$

of solutions $\omega_* = \omega_*(K_\infty)$, $k_* = k_*(K_\infty)$ that are in full accordance with the scalings (3.1a,b) and additional affine transformations (3.8a,b) of time and longitudinal coordinate for the subsonic boundary layer. Thus, the match with the conventional triple-deck theory is attainable in this case. In other words, the mixed time-space derivative $\partial^2\varphi_1/\partial t\partial x$ in (3.30f) becomes indefinitely small as $K_\infty \to -\infty$.

On the contrary, in the limit $K_\infty \to \infty$ the asymptotic representation of solutions to (3.34a,b) reads

$$\omega_* \to -\frac{(-\omega'_*)^3}{k'^{\,2}_*}\frac{1}{K_\infty^2}\left[1 - 2\left(\frac{-\omega'_*}{k'_*}\right)^8 \frac{1}{K_\infty^9} + \cdots\right], \tag{3.36a}$$

$$(3.36b) \qquad k_* \to \frac{(-\omega_*')^3}{k_*'^2}\frac{1}{K_\infty^3}\left[1 - 3\left(\frac{-\omega_*'}{k_*'}\right)^8 \frac{1}{K_\infty^9} + \cdots\right],$$

with the consequence that neutral oscillations do exist for sufficiently large values of the similarity parameter K_∞, i.e., for $M_\infty^2 - 1 \gg R^{-1/9}$. Moreover, as it is seen from (3.36a,b) the critical frequency $\omega_* = \omega_*(K_\infty)$ and wavenumber $k_* = k_*(K_\infty)$ are approaching zero very fast as $K_\infty \to \infty$. As a matter of principle these findings settle the main issue showing how the triple-deck theory of subsonic boundary layer instability can be extended to provide a passage through the threshold value $M_\infty = 1$ and then cover a range of moderate supersonic velocities.

A very simple physical explanation of the phenomenon at hand relies on signal speed and shape considerations. Let us write down the equation

$$(3.37) \qquad \frac{\partial x}{\partial t} + \left(\frac{\partial x}{\partial y}\right)^2 = K_\infty$$

of the characteristic surfaces of (3.30f) taken in the form $x = x(t, y)$. Its solution

$$(3.38) \qquad x - x_0 = K_\infty(t - t_0) + \frac{y^2}{4(t - t_0)}$$

determines the front of a signal emitted at the initial instant $t = t_0$ by a point source in $x = x_0$: this is a parabola with the nose curvature tending to infinity as $t \to t_0$. According to (3.38) external disturbances sweep downstream instantly, whereas they propagate very slowly upstream with respect to an observer at rest. More precisely, the upstream propagation speed of the signal is comparable by an order of magnitude with the phase velocity of the Tollmien-Schlichting waves. As a consequence external pulses accumulate locally and become capable of exerting strong unsteady influence on the vorticity pulsations within the viscous wall sublayer. Obviously, this cumulative effect proves to be negligible when $K_\infty \to -\infty$ due to the rapid increase of the upstream propagation speed relative to an observer at rest.

Further analysis involves calculations to ascertain in more detail the behavior of dispersion curves in the complex ω-plane. As expected, only the first root $\Omega_1 = \Omega_1(k)$ of the dispersion relation (3.33) is associated with the unstable mode. All other roots $\Omega_j = \Omega_j(k)$, $j = 2, 3$ ... give oscillations that decay in time. The critical values ω_* and k_* of the frequency and wavenumber are fixed by the point where the first dispersion curve crosses the real axis. As it is seen from Fig. 3.5, representing a plot of the two dispersion curves ω_1 for $K_\infty = \mp 1$, their behavior is similar in general to the behavior of the same curve in Fig. 3.2 for the subsonic boundary layer if the similarity parameter takes on moderate values. At the same time, neither curve in Fig. 3.5 bears any resemblance to those in Fig. 3.3; thus the new approach capability

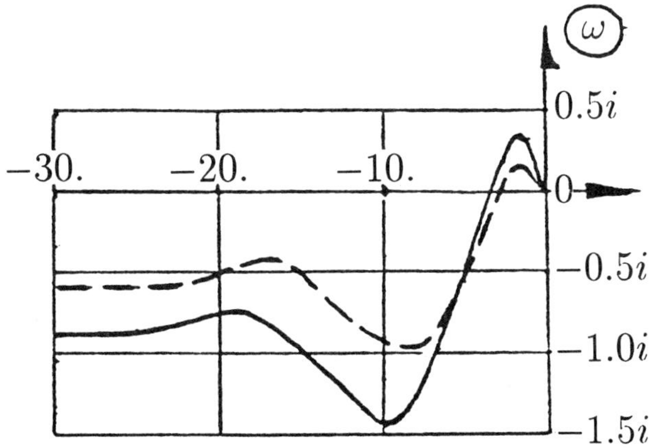

FIG. 3.5. *The traces of the first root $\omega = \omega_1(k)$ in the complex ω-plane with k running through real positive values for a transonic boundary layer. The right-hand side of the dispersion relation* (3.33) *is given by* (3.27c) *with λ as in* (3.26b,c)*:* ——— $K_\infty = -1$; — — — $K_\infty = 1$.

of predicting instability properties of a moderately supersonic boundary layer appears to be evident in full measure.

Our concern now is with estimating the Mach number range where the theory developed is valid. Let ν^* be the frequency in the original system of dimensional units. Then a reduced frequency

$$\nu = \frac{\nu^* L^*}{U_\infty^*}\epsilon^2 \frac{C^{1/4}}{\tau_w^{3/2}\omega_*'}\frac{T_w^*}{T_\infty^*} \tag{3.39}$$

turns out to be the most suitable entity insofar as its critical value ν_* for neutral disturbances propagating in the subsonic boundary layer is simply

$$\nu_* = (1 - M_\infty^2)^{1/4}. \tag{3.40}$$

Owing to the normalized time definition (3.28a) it is nothing more than the first term from the asymptotic expansion (3.35a) for ω_* in the limit as $K_\infty \to -\infty$, though in general the two frequencies under consideration are connected by

$$\nu_* = \Delta\omega_*(K_\infty)/\omega_*', \qquad K_\infty = (M_\infty^2 - 1)/\Delta^4, \tag{3.41a, b}$$

where

$$\Delta = [4\epsilon^2 C^{1/4}\tau_w^{1/2}(T_w^*/T_\infty^*)]^{1/9}. \tag{3.41c}$$

On the other hand, the limit $K_\infty \to -\infty$ is attainable only provided the difference $1 - M_\infty^2$ is positive and fixed as $\epsilon \to 0$. However, the values of the Reynolds number, even in the range corresponding to laminar-turbulent transition, give a rough estimation $\epsilon \approx 0.2$. In a boundary layer on a thermally insulated plate T_w^*/T_∞^* obeys the law (see, for example, [17])

$$\frac{T_w^*}{T_\infty^*} = 1 + \frac{\gamma - 1}{2} M_\infty^2, \tag{3.42}$$

γ being the ratio of specific heats. Hence $T_w^*/T_\infty^* \to (\gamma+1)/2$ as $M_\infty \to 1$. Let us assume that the temperature of the surrounding air is 293°K; then $\gamma = 1.4$ and $C \approx 0.79$ which results in $\Delta \approx 0.78$. As a consequence the similarity parameter K_∞ is running through the values $-2.73 \le K_\infty \le 8.19$ for Mach numbers in the range $0 \le M_\infty \le 2$. Thus, the limit $K_\infty \to -\infty$ is hardly approximated under real conditions.

The curves drawn in accordance with (3.40) and (3.41a–c) are labeled in Fig. 3.6 with the numbers 1 and 2, respectively. They are approaching each other when the Mach number decreases; however, the relative distance separating them is still as large as 0.14 due to the finite value of ϵ and, consequently, K_∞ even if $M_\infty = 0$. So the critical frequency of eigenoscillations reaches its subsonic limit very slowly. On the contrary, the asymptotic behavior $\nu_* = \Delta^9(\omega_*'/k_*')^2/(M_\infty^2 - 1)^2$ as prescribed by (3.36a) for the supersonic boundary layer and illustrated with the dotted line in Fig. 3.6 sets in rather quickly. This line merges in the curve 2 already at $M_\infty = 1.5$.

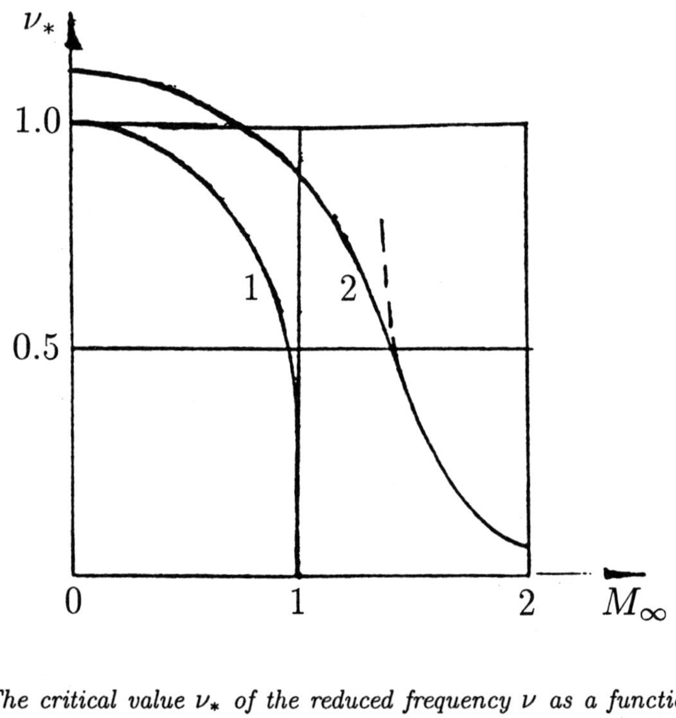

FIG. 3.6. *The critical value ν_* of the reduced frequency ν as a function of the Mach number M_∞: 1, a plot of (3.40); 2, a plot of (3.41a–c); — — —, an asymptote given by (3.36a).*

3.5. Cauchy Problem

The amplitude growth rate of the Tollmien-Schlichting waves is tending to $\sqrt{2}/2$ following asymptotics

$$\omega_1 = -k^2 - \sqrt{\frac{2}{2}}(1+i) + \cdots \tag{3.43}$$

of the first root of the dispersion relation (3.12a–c) with the right-hand side given by (3.13) for the subsonic boundary layer as $k \to \infty$ ([22], [13]). The aforementioned finite limit ensures the Cauchy problem to be well-posed. For this reason the triple-deck theory accepts unrealistic eigenmodes with large wavenumbers that are exponentially increasing with time.

The situation changes drastically when we consider stability properties of the transonic boundary layer flow and substitute $Q(\omega, k) = (ik)^{1/3}k^2/\lambda$ prescribed by (3.27c) for the right-hand side of (3.13). In this case the asymptotic expansion for ω_1 reads [11]

$$\omega_1 = -\left(k^{5/3} + \frac{1}{3}K_\infty k + \frac{1}{9}K_\infty^2 k^{1/3}\right) - \frac{\sqrt{2}}{3}(1+i)k^{1/6} + \ldots \tag{3.44}$$

if K_∞ is fixed and $k \to \infty$, provoking the amplitude growth rate to enhance as fast as $\sqrt{2}k^{1/6}/3$. Thus, the term $\partial^2\varphi_1/\partial t\partial x$ in (3.30f) produces the necessary destabilizing effect for moderate values of the wavenumber k in the upper deck, but the same effect becomes too strong as $k \to \infty$. Therefore we need to distinguish between two limit passages

$$\text{(i)} \quad \epsilon \to 0, \quad K_\infty \to \infty, \quad k \text{ is fixed;} \tag{3.45a}$$

$$\text{(ii)} \quad \epsilon \to 0, \quad k \to \infty, \quad K_\infty \text{ is fixed.} \tag{3.45b}$$

The first one allows the critical values ω_* and k_* of the frequency and wavenumber to be evaluated in the forms (3.36a) and (3.36b), respectively. The second limit process leads to the Cauchy problem that turns out to be ill-posed. However, in considering any physical problem we deal with the second type of the limits for the Reynolds number R (and ϵ) along with the similarity parameter K_∞ are given by conditions at infinity and on the body surface.

To make constraints of the both limit processes under consideration somewhat weaker and bring also more to light the properties of eigenoscillations in the supersonic boundary layer let us normalize the wavenumber in the same manner $k = k_0 K_\infty^{-3/8}$, $k_0 = \text{const}$ as is the case for a subsonic mainstream and suppose that $K_\infty \to \infty$. Actually, this is an intermediate limit to connect (3.45a) with (3.45b) because the similarity parameter K_∞ and the wavenumber k are allowed to take on indefinitely large values simultaneously. Then an estimation $\omega = O(K_\infty^{11/8})$ supersedes that given by (3.35a) and implies the sharp increase in the pulsation frequency when passing through the threshold value $M_\infty = 1$ of the Mach number. The resulting asymptotic representation becomes

$$\omega_1 = -(k_0 K_\infty^{11/8} + k_0^3 K_\infty^{-7/8}) - \sqrt{2}(1+i)k_0^{5/2}K_\infty^{-41/16} + ... \tag{3.46}$$

It is indicative of the amplitude growth rate attenuation even for disturbances with large wavenumbers if the similarity parameter $K_\infty \to \infty$.

To overcome the difficulty that arises in the limit process (3.45b) due to the ill-posed Cauchy problem the basic assumptions of the triple-deck theory need to be revised. We may choose two alternative ways for improving the asymptotic predictions of the short-length wave evolution. The first one is to rescale both the independent variables and unknown functions when tackling (3.45b) in order to derive a new set of simplified equations. Their solution should be matched further on to the solution of (3.30a–h) with the similarity parameter K_∞ fixed by (3.31). An alternative approach is to include into analysis the main second-order terms evaluated within the conventional version of the triple deck. Basically, this is an account for the finite values of the Reynolds number leading to a system of composite equations. The second way is more preferable since it provides a unified treatment of stability characteristics for moderate as well as large values of the wavenumber.

To begin with, let us reconsider the disturbance pattern in the main deck II in Fig. 3.1; it embraces most of the boundary layer with the transverse coordinate y_2 measuring the scaled distance from the wall in the usual manner adopted in the classic theory by Prandtl. The excess pressure p_2 can be written in this region as

$$p_2 = p_{21}(t,x) + \epsilon p_{22}(t,x,y_2) + ... \tag{3.47}$$

where the second-order correction p_{22} is to be expressed in terms of the first-order displacement thickness $-A_1(t,x)$ through ([12])

$$\frac{\partial p_{22}}{\partial y_2} = -|\delta'|^{7/8} R_0(y_2) U_0^2(y_2) \frac{\partial^2 A_1}{\partial x^2}. \tag{3.48}$$

The functions R_0 and U_0 are the density and velocity distributions across the undisturbed boundary layer, respectively. It is convenient to define the Mach number distribution M_0 by means of

$$M_0^2(y_2) = M_\infty^2 R_0(y_2) U_0^2(y_2) \tag{3.49}$$

in order to reduce an expression that results from integrating (3.48) to the form

$$p_{22} = p_{22}(t,x,0) + |\delta'|^{7/8}\left[y_2 - \int_0^{y_2} \frac{M_\infty^2 - M_0^2(Y)}{M_\infty^2} dY\right] \frac{\partial^2 A_1}{\partial x^2}. \tag{3.50}$$

Let $p_{12}(t, x, y_1)$ and $p_{32}(t, x)$ be the second-order pressures in the upper and lower decks, respectively. The matching conditions require that

(3.51)
$$p_{32}(t, x) = p_{22}(t, x, 0) = p_{12}(t, x, 0) + |\delta'|^{7/8} \frac{\partial^2 A_1}{\partial x^2} \int_0^\infty \frac{M_\infty^2 - M_0^2(Y)}{M_\infty^2} dY.$$

In the case of the boundary layer flow past a thermally insulated flat plate considered throughout this paper, the magnitude of the integral in the right-hand side of (3.51) is available in [17], namely

$$\int_0^\infty \frac{M_\infty^2 - M_0^2(Y)}{M_\infty^2} dY = \Delta_1 (2C)^{1/2} \frac{T_w^*}{T_\infty^*}, \qquad \Delta_1 = 1.686. \tag{3.52}$$

The crucial point is to evaluate the second-order pressure $p_{12}(t, x, 0)$ at the bottom of the outer region. This is a complicated problem in general; however, it can be simplified when studying steady separation or any other kind of time-independent phenomena. Then a composite representation

$$p = -\frac{\partial \varphi_1}{\partial x} + \epsilon |\delta'|^{7/8} \Delta_1 (2C)^{1/2} \frac{T_w^*}{T_\infty^*} \frac{\partial^2 A}{\partial x^2} \qquad \text{at} \quad y_1 = 0 \tag{3.53}$$

holds true instead of (3.21a) for matching the total pressure within the boundary layer to that generated in the potential field of both transonic regimes, if notation accepted in §3.2 is employed. Introducing here variables (3.28b) and (3.29c–e) of §3.4 with $|\delta'| = \epsilon^{8/9}$ we arrive at the final form

$$p = -\frac{\partial \varphi_1}{\partial x} + \epsilon^{16/9} D \frac{\partial^2 A}{\partial x^2} \qquad \text{at} \quad y_1 = 0 \tag{3.54}$$

of the boundary condition to be substituted for (3.30g) where the constant D depends on the reference quantities C, τ_w and T_w^*/T_∞^* as follows

$$D = 2^{23/18} \Delta_1 C^{2/9} \tau_w^{13/9} (T_w^*/T_\infty^*)^{-1/9}. \tag{3.55}$$

Since $-A$ is a normalized displacement thickness, the second derivative $\partial^2 A/\partial x^2$ can be approximately treated as the streamline curvature that creates the centrifugal forces in the flow field. Then, as it is obvious from (3.54), the less the distance is within which the gas particles deviate from their original directions the more significant the contribution of the latter forces to the pressure variations becomes. This conclusion lends a strong support to the above statement concerning the growing role of the normal pressure gradient for large values of the wavenumber.

Strictly speaking the composite representation (3.53) holds for steady flows. So, this condition is not too much restrictive for the first transonic regime considered in §3.3 because it is mainly aimed at elucidating separation and other essentially nonlinear time-independent phenomena. On the contrary, the loss of the boundary layer stability inherent in the second transonic regime is governed by the unsteady interaction process that involves two different systems of waves propagating in the potential region and the wall sublayer. Nevertheless, we can use (3.53) in order to bring to light how the asymptotics (3.44) alter if the stabilizing effect due to the normal pressure gradient is included into analysis. The modified asymptotics of the complex frequency for large wavenumbers make, in turn, the Cauchy problem well-posed as it will be shown in what follows.

With (3.54) and (3.55) taken for expressing pressure variations across the viscous wall sublayer the dispersion relation arises again in the form (3.33) with the right-hand side

$$Q(\omega, k) = (ik)^{1/3}k^2\left(\frac{1}{\lambda} - \epsilon_0\right), \qquad \epsilon_0 = \epsilon^{16/9}D, \tag{3.56a, b}$$

and λ being as in (3.26b,c). In the limit $\mid \Omega \mid \rightarrow \infty$ we have in the first approximation

$$-\omega_1(-\omega_1 k - K_\infty k^2)^{1/2} = k^3[1 - \epsilon_0(-\omega_1 k - K_\infty k^2)^{1/2}] \tag{3.57}$$

assuming, to be specific, that $k > 0$. To simplify the following analysis let us employ a quantity

$$a = \epsilon_0 \lambda = \epsilon_0(-\omega_1 k - K_\infty k^2)^{1/2} \tag{3.58}$$

instead of the frequency ω_1. On introducing a parameter $K = \epsilon_0^{1/2} K_\infty$ the approximate form (3.57) of the dispersion relation reduces to the cubic equation

$$a^3 + (Kd^{1/2} + d)a - d = 0 \tag{3.59}$$

for the new variable a considered to be a function of the product $d = \epsilon_0^3 k^4$. So, the singular wavenumber range is estimated as $k = O(\epsilon_0^{-3/4}) = O(\epsilon^{-4/3})$. An estimation of the corresponding frequency range is derivable from (3.58) and reads $\omega_1 = O(\epsilon_0^{-5/4}) = O(\epsilon^{-20/9})$ since $a = O(1)$ when $d = O(1)$.

Let us come back to the limit process (3.45b) governing the Cauchy problem properties. Its constraints can be met if both $\epsilon \rightarrow 0$ and $k \rightarrow \infty$ are chosen from the singular range provided that the similarity parameter K_∞ is kept fixed. Then, by definition, $K \rightarrow 0$ and the second term in the left-hand side of (3.59) becomes negligibly small, hence

$$a^3 + da - d = 0. \tag{3.60}$$

The latter equation has three real roots for any $d > 0$, only one of which pertains to the eigenmode analysis at hand, namely

$$a = \left[\frac{1}{2}d + \left(\frac{1}{4}d^2 + \frac{1}{27}d^3\right)^{1/2}\right]^{1/3} + \left[\frac{1}{2}d - \left(\frac{1}{4}d^2 + \frac{1}{27}d^3\right)^{1/2}\right]^{1/3}. \tag{3.61}$$

Now an asymptotic expansion for ω_1 appears in the form

$$\omega_1 = -(c_0 k^{5/3} + c_1 k + c_2 k^{1/3}) - b(1+i)k^{1/6} + ..., \tag{3.62}$$

similar to (3.44) where c_0 is expressed in terms of d and a as

$$c_0 = d^{-2/3}a^2. \tag{3.63}$$

The subsequent constants are

$$c_1 = \frac{K_\infty}{1+2c_0^{3/2}}, \qquad c_2 = \frac{3c_0^2 K_\infty^2}{(1+2c_0^{3/2})^3}, \qquad b = \frac{\sqrt{2}\, c_0}{1+2c_0^{3/2}}. \tag{3.64a, b, c}$$

If $d \ll 1$ we conclude from (3.61) that $a \to d^{1/3}$ with the result $c_0 \to 1$ following from (3.63). The asymptotic expansion (3.44) is retrieved in this case, which might be regarded as giving a starting point for the dispersion curve location in the complex ω-plane for sufficiently large values of the wavenumber. However, (3.61) yields $a \to 1$ when $d \gg 1$; therefore, the both constants of concern, c_0 and b, are tending to zero in accordance with (3.63) and (3.64c), respectively.

Thus, the normal pressure gradient exerts necessary stabilizing influence on the disturbance development making the amplitude growth rate to drastically decrease as it has been supposed at the beginning of this section. Certainly, it is impossible to proceed too far with $d \gg 1$ in calculating the dispersion curve shape because $\Omega = O(\epsilon_0^{-3/4} d^{-5/12}) = O(\epsilon^{-4/3} d^{-5/12})$ ceases to be large in the range $d = O(\epsilon_0^{-9/5}) = O(\epsilon^{-16/5})$ violating the basic assumption $|\, \Omega \,| \to \infty$ for (3.57) to be valid.

3.6. Neutral Oscillations

To continue the above analysis we ought to address a precise form of the dispersion relation and treat it numerically. However, neutral oscillations represent an exception because their existence can be established without tedious calculations. The dispersion relation conveniently written for this purpose as

$$\frac{\omega_*}{k_*^{2/3}} = \frac{\omega_*'}{k_*'^{\,2/3}}, \qquad \frac{k_*^{7/3}\,[1-\epsilon_0(-\omega_* k_* - K_\infty k_*^2)^{1/2}]}{(-\omega_* k_* - K_\infty k_*^2)^{1/2}} = k_*'^{\,4/3} \tag{3.65a, b}$$

contains, in contrast to (3.34a,b), an additional term proportional to the small parameter ϵ_0. Crucial simplifications come again from introducing new variables

$$d_* = -\frac{\omega_*'}{k_*'^{\,2}}\epsilon_0^3 k_*^4, \qquad a_* = \epsilon_0 \lambda_* = \epsilon_0(-\omega_* k_* - K_\infty k_*^2)^{1/2}, \tag{3.66a, b}$$

and a parameter

$$K_* = \left(-\frac{k_*'^{\,2}}{\omega_*'}\right)^{1/2} \epsilon_0^{1/2} K_\infty \tag{3.67}$$

similar to those exploited in the previous section. Here $\omega'_* = -2.298$ and $k'_* = 1.0005$ designate, as usual, the critical values of the frequency and wavenumber inherent in disturbances propagating through the subsonic boundary layer. The quantity a_* considered as a function of d_* satisfies the cubic equation (3.59) with K_* substituted for K. Thus, (3.59) is equally applied for developing an asymptotic expansion defining the frequency when the wavenumber is tending to infinity and for evaluating the critical values of the same quantities for neutral oscillations. Nevertheless, there is an essential distinction between the two cases: the former is fixed solely by (3.59), to make the latter complete we should regard either (3.66a) or (3.66b) as an supplementary constraint to be imposed on a solution of the cubic equation at hand. It is worth stressing that the critical values sought are not subject to any a priori limitations: they can be both moderate and large (for instance, lie within the singular range).

Let us assume first that the critical frequency and wavenumber are both of $O(1)$ in magnitude, then $d_* \to 0$ and $K_* \to 0$ with $\epsilon_0 \to 0$ in accordance with (3.66a) and (3.67), respectively. Insofar as the product $| K_* | d_*^{-1/2} \to \infty$, the third term in the left-hand side of (3.59) becomes negligibly small, therefore

$$(3.68) \qquad a_*^3 + K_* d_*^{1/2} a_* - d_* = 0$$

supersedes (3.60). Being rewritten in terms of the original quantities k_*, λ_*, and K_∞ the approximate dispersion relation (3.68) is nothing more than the similar relation that holds in the absence of the normal pressure gradient. The small parameter ϵ_0 drops out of the final results in line with a set of governing equations and boundary conditions where it does not enter in this case. The asymptotics for subsonic and supersonic mainstream are easily retrieved as a consequence. Indeed, if $M_\infty < 1$ and the parameter $K_* \ll -1$ then (3.68) leads to $a_* \to | K_* |^{1/2} d_*^{1/4}$ resulting in (3.35a,b). On the contrary, with $M_\infty > 1$ and $K_* \gg 1$ a pertinent root

(3.69)
$$a_* = \left[\frac{1}{2}d_* + \left(\frac{1}{4}d_*^2 + \frac{1}{27}K_*^3 d_*^{3/2}\right)^{1/2}\right]^{1/3} + \left[\frac{1}{2}d_* - \left(\frac{1}{4}d_*^2 + \frac{1}{27}K_*^3 d_*^{3/2}\right)^{1/2}\right]^{1/3}$$

of (3.68) yields $a_* \to K_*^{-1} d_*^{1/2}$ whence (3.36a,b) are derivable. Certainly, for a sonic flow at infinity with $M_\infty = 1$ and $K_* = 0$ the critical values

$$(3.70\text{a, b}) \qquad \omega_* = -\left(-\frac{\omega_*'^{11}}{k_*'^2}\right)^{1/9}, \qquad k_* = (-\omega_*' k_*'^2)^{1/3},$$

are both of $O(1)$ in magnitude. Thus, the first pair of the neutral oscillation frequency and wavenumber coincides with that predicted by the conventional version of the triple-deck theory, including the limit cases of moderately subsonic and supersonic boundary layers.

Let us inquire now into the possibility for a second pair to exist that is, obviously, beyond the scope of the conventional triple-deck version. Pressure variations across the main part of the boundary layer should be necessarily taken into account for this purpose. To be specific, let the reduced frequency $d_* \to |K_*|^{-2}$ with $K_* = -|K_*|$ and $a_* \to 1$ in order to satisfy the dispersion relation (3.59). So, only subsonic velocities in the mainstream are admitted now. Allowing for small deviations from these quantities we put

(3.71a, b) $$d_* = |K_*|^{-2}(1+\delta_d), \qquad a_* = 1+\delta_a,$$

and derive a linear dependence

(3.72) $$\delta_d = 2(2+|K_*|^{-2})\delta_a$$

to connect δ_d with δ_a. It remains still to go back to the initial equations (3.65a,b), the second of which gives

(3.73) $$\delta_a = -k_*'^{4/3}[\epsilon_0 |K_\infty|^{7/8}]^{4/3}.$$

For ensuring δ_a to remain less than $O(1)$ we need to impose the following constraint

(3.74) $$\epsilon_0 |K_\infty|^{7/8} \ll 1$$

on ϵ_0 and $|K_\infty|$ which in terms of the two original small parameters ϵ and $|M_\infty^2 - 1|$ reads

(3.75) $$\epsilon |M_\infty^2 - 1|^{7/8} \ll 1.$$

The same condition that has been actually used for the second term could be introduced into the right-hand side of (3.53) prescribing the excess pressure. In conclusion, (3.65a) is expressed as

(3.76) $$4\delta_a - \delta_d = 2\left(-\frac{\omega_*'}{k_*'^{2/3}}\right)[\epsilon_0 |K_\infty|^{-5/2}]^{1/3}.$$

Combining (3.72) and (3.76) we arrive again at (3.73), nevertheless (3.72) shows that another constraint

(3.77) $$\epsilon_0 |K_\infty|^{-5/2} \ll 1$$

should be met since both δ_a and δ_d are small by definition. In terms of original parameters, (3.77) takes on the form

(3.78) $$\epsilon |M_\infty^2 - 1|^{-5/8} \ll 1.$$

If $|K_\infty| = O(1)$ the both conditions (3.75) and (3.78) indicate simply that $\epsilon \to 0$ in compliance with the basic assumptions exposed in §3.1. In the limit $K_\infty \ll -1$ associated with the subsonic boundary layer only the first constraint (3.74) is meaningful while (3.77) plays no part. On the contrary, in the transonic range proper when $K_\infty \to 0-$ a decisive role belongs to the second constraint (3.77) and (3.74) becomes of no importance. Moreover, (3.78) allows the results of the analysis developed to be extended as far as the second transonic regime under consideration sets in overlapping with the first one defined by the similarity parameter (3.25). Omitting in the right-hand side of (3.71a) the correction δ_d we conclude from the leading terms

$$(3.79\text{a, b}) \qquad k_* = [\epsilon_0 \mid K_\infty \mid^{1/2}]^{-1}, \qquad \omega_* = \frac{\omega'_*}{k'^{2/3}_*}[\epsilon_0 \mid K_\infty \mid^{1/2}]^{-2/3},$$

of the critical values sought for that $k_* \to \infty$, $\mid \omega_* \mid \to \infty$ if $K_\infty \to 0-$ and ϵ is small but fixed. In the limit $K_\infty \ll -1$ two other expressions

(3.80a, b)

$$k_* = \mid K_\infty \mid^{3/8} [\epsilon_0 \mid K_\infty \mid^{7/8}]^{-1}, \qquad \omega_* = \frac{\omega'_*}{k'^{2/3}_*} \mid K_\infty \mid^{1/4} [\epsilon_0 \mid K_\infty \mid^{7/8}]^{-2/3},$$

are preferable to show that k_* and $\mid \omega_* \mid$ are large even on scales adopted in the subsonic triple-deck theory. So, accounting for the normal pressure gradient reveals the second pair of the critical frequency and wavenumber to exist provided that $M_\infty < 1$ in the mainstream. The quantities defined by (3.79a,b) or (3.80a,b) are distinct from those associated with the upper branch of the neutral stability curve. This phenomenon has not been known before in the standard hydrodynamic stability theory (see, for example, [3]). The new critical quantities become indefinitely large with $\epsilon \to 0$ and $R \to \infty$, if M_∞ is kept fixed. They cease to come into play for the external flow with strictly sonic velocity even though ϵ is considered to be fixed (but, certainly, small). The second pair of the critical frequency and wavenumber disappears in disturbances propagating through the supersonic boundary layer with $M_\infty > 1$.

3.7. Acknowledgments

This research stems from collaboration with Dr. I. V. Savenkov, Computer Center, the USSR Academy of Sciences, and references therein reflect his participation at the earlier stage of the work. More recent results have been obtained during the author's stay at the Dept. of Mathematical Sciences of the University of Delaware, which allowed to complete investigations undertaken later, at the Dept. of Mathematical Sciences of Rensselaer Polytechnic Institute. It is my pleasant duty to express sincere thanks to Professors L. P. Cook and J. D. Cole for numerous discussions and encouraging comments. The author acknowledges gratefully also many improvements introduced by Dr. G. Schleiniger. The final stage of this study was carried out with the support of the Alfred P. Sloan Foundation and the Air Force Office of Scientific Research under grant AFOSR 88-0037.

References

[1] R. J. Bodony and A. Kluwick, *Freely interacting transonic boundary layers*, Phys. Fluids, 20 (1977), pp. 1432–1437.

[2] J. D. Cole and L. P. Cook, *Transonic Aerodynamics*, North-Holland, Amsterdam, 1986.

[3] P. G. Drazin and W. H. Reid, *Hydrodynamic Stability*, Cambridge University Press, Cambridge, 1981.

[4] M. J. Lighthill, *On boundary layers and upstream influence* II, *Supersonic flows without separation*, Proc. Roy. Soc. London, Ser. A, 217 (1953), pp. 478–507.

[5] C. C. Lin, E. Reissner, and H. S. Tsien, *On two-dimensional non-steady motion of a slender body in a compressible fluid*, J. Math. Phys., 27 (1948), pp. 220–231.

[6] A. F. Messiter, *Boundary-layer flow near the trailing edge of a flat plate*, SIAM J. Appl. Math., 18 (1970), pp. 241–257.

[7] A. F. Messiter, A. Feo, and R.E. Melnik, *Shock wave strength for separation of a laminar boundary layer at transonic speeds*, AIAA J., 9 (1971), p. 1197.

[8] V. Ya. Neiland, *Contribution to the theory of separation of a laminar boundary layer in a supersonic stream*, Izv. Akad. Nauk SSSR, Mekh. Zhidk. Gaza, 4 (1969), pp. 53–57 (in Russian).

[9] O. S. Ryzhov, *On an unsteady boundary layer with the self-induced pressure in the transonic velocity range of an external stream*, Dokl. Akad. Nauk SSSR, 236 (1977), pp. 1091–1094 (in Russian).

[10] ——, *Asymptotic methods in transonic flow theory*. In Symposium Transsonicum III, J. Zierep and H. Oertel, eds., Springer-Verlag, Berlin, 1989, pp. 143–155.

[11] O. S. Ryzhov and I. V. Savenkov, *On the boundary-layer stability in the transonic velocity range of an external stream*, Zh. Prikl. Mekh. i Techn. Fiz., 2 (1990), pp. 65–71 (in Russian).

[12] O. S. Ryzhov and E. D. Terent'ev, *On an unsteady boundary layer with the self-induced pressure*, Prikl. Mat. Mekh., 41 (1977), pp. 1007–1023 (in Russian).

[13] ——, *On some properties of vortex spots in a boundary layer on a flat plate*, Dokl. Akad. Nauk SSSR, 276 (1984), pp. 551–575 (in Russian).

[14] ——, *On a transient regime that is characteristic of the launch of a vibrator in a subsonic boundary layer on a flat plate*, Prikl. Mat. Mekh., 50 (1986), pp. 974–986 (in Russian).

[15] O. S. Ryzhov and V. I. Zhuk, *Internal waves in the boundary layer with the self-induced pressure*, J. Mécanique, 19 (1980), pp. 561–580.

[16] F. T. Smith, *On the non-parallel flow stability of the Blasius boundary layer*, Proc. Roy. Soc. London, Ser. A, 366 (1979), pp. 91–109.

[17] K. Stewartson, *The Theory of Laminar Boundary Layers in Compressible Fluids*, Clarendon Press, Oxford, 1964.

[18] ——, *On the flow near the trailing edge of a flat plate* II, Mathematika, 16 (1969), pp. 106–121.

[19] K. Stewartson and P. G. Williams, *Self-induced separation*, Proc. Roy. Soc. London, Ser. A, 312 (1969), pp. 181–206.

[20] E.D. Terent'ev, *The linear problem on vibrator in a subsonic boundary layer*, Prikl. Mat. Meckh., 45 (1981), pp. 1049–1055 (in Russian).

[21] V. I. Zhuk and O. S. Ryzhov, *Free interaction and stability of a boundary layer in incompressible fluid*, Dokl. Akad. Nauk SSSR, 253 (1980), pp. 1326–1329 (in Russian).

[22] ——, *Asymptotic behavior of solutions of the Orr–Sommerfeld equation that yield unstable oscillations at large Reynolds numbers*, Dokl. Akad Nauk SSSR, 268 (1983), pp. 1328–1332 (in Russian).

Chapter 4

Choked Wind-Tunnel Flow: Asymptotics and Numerics

L. Pamela Cook and Gilberto Schleiniger

4.1. Introduction

Airfoils and wings are routinely tested as scaled models in wind-tunnels. It is imperative therefore to understand wind-tunnel wall interference, that is, the effect the wind-tunnel walls have on the flow, so that the free flight characteristics of the wing can be determined. In this paper we review results specifically for choked solid wall wind-tunnel flows. Although choked flows are not of themselves of wide practical interest they are important as limiting cases and hence for an understanding of nearly choked flows. In particular the order of magnitude of the wind-tunnel wall interference changes significantly as the supersonic region gets near the tunnel walls, that is, as the flow approaches choking. This paper discusses both 2D and 3D as well as both lifting and nonlifting choked flows in solid wall wind-tunnels.

Choked flow is characterized by the fact that the sonic surface, originating on the wing, extends to the walls. The characteristic surface that meets the wall at the sonic surface is the limiting characteristic surface. This surface (or correspondingly the two limiting characteristics in 2D flow) divides the flow into an upstream region and a downstream region. Any portion of the latter can have no effect on the former. Clearly to maintain choking conditions a special balance must exist between the upstream velocity U, the wind-tunnel height h, and the body thickness δ (and its shape). The existing analyses, as with experimental testing, are specifically concerned with thin wings in wide wind-tunnels. For those conditions one expects the upstream (choking) Mach numbers to be close to sonic (one) and the flow, at least near the body, to be close to sonic. Thus the analysis has some commonality with that of the *stabilization law* [3], [7]. The asymptotic analysis provides orders of magnitude of the correction to sonic free flow as well as of the dependence of the choking Mach number on the wind-tunnel height. The actual numerical values are of interest in both cases and to obtain those we need to solve the resulting nonlinear boundary value problems numerically.

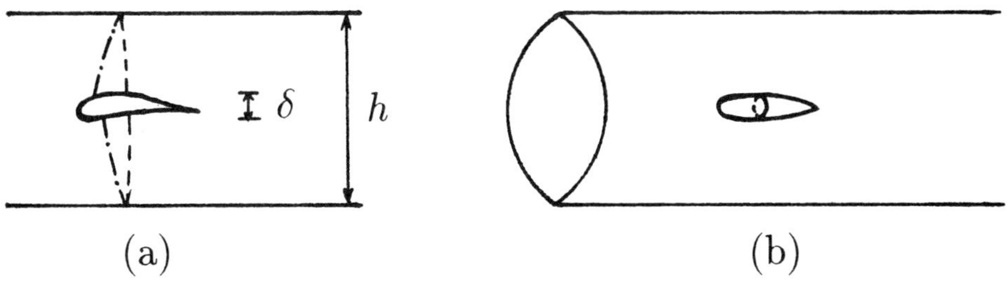

FIG. 4.1. *Choked wind-tunnel flow: (a) 2D, (b) 3D, - · - · - sonic line, - - - characteristic.*

In this paper we review results for these flows, in particular earlier work by Guderley [9] and by Marschner [12], Morioka [13], Sevost'ianov [14], and Cole and Cook [2], [3]. Some discussion of slower flows is also given as in Malmuth and Cole [11] and in Malmuth's paper in this volume. We begin by summarizing the results of Cole and Cook, which are within the context of transonic small disturbance theory and then we compare these results with earlier results and discuss some work in progress. For a review of issues in the application of transonic flow theory to wind-tunnel interference at sonic and subsonic speeds (calibration of the test section, etc.) see Berndt [1].

4.2. Small Disturbance Flow

Consider a profile given by

$$B\left(x, \frac{y}{\delta}, \frac{z}{\delta}\right) = y - \delta F_{u,\ell}\left(x, \frac{z}{\delta}\right) = 0, \quad 0 \leq x \leq 1$$

for a slender body and

$$S(x, y) = y - \delta F_{u,\ell}(x) = 0, \quad 0 \leq x \leq 1$$

for an airfoil, where $\delta << 1$ is the thickness ratio, u, ℓ refer to the upper and lower surfaces, respectively, and the length scale is the airfoil chord. Within the framework of transonic small disturbance theory we consider the limit $\delta \longrightarrow 0$ as the scaled transverse coordinates $\tilde{y} = \epsilon(\delta)\, y$, $\tilde{z} = \epsilon(\delta)\, z$ as well as the transonic similarity parameter $K_T = (1 - M_\infty^2)/\epsilon^2$ and the scaled height $H = \epsilon\, h$ are kept fixed [4]. Here, for 2D flow $\epsilon = \delta^{1/3}$, and for slender body flow $\epsilon = \delta$. The scaling brings the slow changes in the transverse direction onto the same scale as the rapid changes in the streamwise direction. Since weak shock waves introduce entropy changes only to the order of ϵ^6 it is sufficient here to consider a potential Φ. The potential can be expanded about the uniform stream so that

$$\Phi(x, y, z; U, \delta, h) = U\{x + \epsilon^2(\delta)\, \phi(x, \tilde{y}, \tilde{z}; K_T, H) + \cdots\}\,.$$

With $\tilde{\nabla}^2 = \partial^2/\partial\tilde{y}^2 + \partial^2/\partial\tilde{z}^2$ for 3D flow, $\tilde{\nabla}^2 = \partial^2/\partial\tilde{y}^2$ for 2D flow, and $\tilde{r}^2 = \tilde{y}^2 + \tilde{z}^2$ for 3D flow, $\tilde{r}^2 = \tilde{y}^2$ for 2D flow, we have that ϕ satisfies the boundary value problem

$$(K_T - (\gamma + 1)\phi_x)\phi_{xx} + \tilde{\nabla}^2\phi = 0 \qquad (4.1)$$

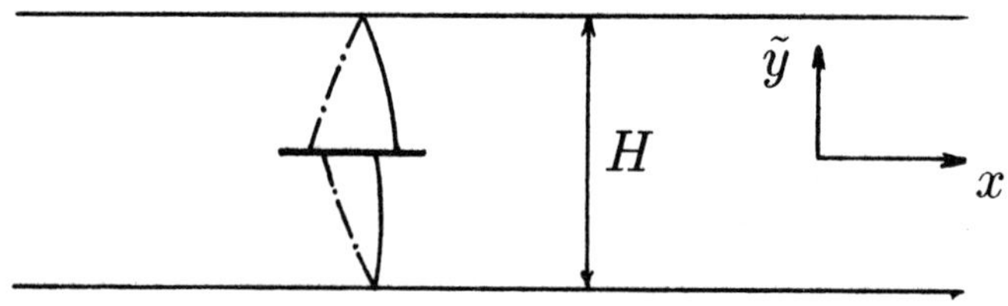

FIG. 4.2. *2D wind-tunnel flow: small disturbance coordinates.*

with

$$\frac{\partial\phi}{\partial n} = 0 \quad \text{at the wall} \quad \text{(solid tunnel walls)}, \tag{4.2}$$

$$\phi_x \to 0 \quad \text{as} \quad x \to -\infty \quad \text{(no disturbance at upstream infinity)}, \tag{4.3}$$

and for 3D slender body flow in an axisymmetric wind-tunnel

$$\tilde{r}\phi_{\tilde{r}} \to \frac{1}{2\pi}\frac{\partial A^*}{\partial x} \quad \text{as} \quad \tilde{r} \to 0, \quad 0 \le x \le 1, \tag{4.4}$$

where

$$A^* = \int_{-z^*_{LE}(x)}^{z^*_{LE}(x)} (F_u - F_\ell)\, d\tilde{z} \tag{4.5}$$

is the normalized cross-section area, and for 2D flow

$$\phi_{\tilde{y}}(x, 0\pm) = F'_{u,\ell}(x), \quad 0 \le x \le 1. \tag{4.6}$$

There are also shock conditions namely

$$[\![\phi]\!]_s = 0, \tag{4.7}$$

and

$$[\![K_T\phi_x - \frac{\gamma+1}{2}\phi_x^2]\!]_s\, [\![\phi_x]\!]_s + [\![\phi_{\tilde{r}}]\!]_s^2 = 0 \tag{4.8}$$

where $[\![\ ]\!]_s$ denotes the jump across the shock and, for 2D flow, the Kutta condition must be satisfied, namely no jump in pressure across the trailing edge

$$[\![\phi_x]\!]_{TE} = 0 \quad \text{at} \quad x \ge 1,\ y = 0. \tag{4.9}$$

Note that, to this order, for a slender body, the flow is axisymmetric. Note that the shocks for a 2D airfoil stabilize (as $M_\infty \to 1$) near the tail, and for axisymmetric flow where $\mathcal{S}' = 0$, with $\mathcal{S} = (1/2\pi)\partial A^*/\partial x$ [5]. We will consider flows for which the shock is behind the limiting characteristic surface and hence it need not be considered for calculations of the front part of the flow. Now (4.1)–(4.6) must be examined to find for a given H which K_T (or vice versa) will give rise to choked flow, and what the correction to the flow is due to the walls.

4.3. Choked Flow—Small Disturbance Theory

For large H ($H \to \infty$) the flow splits naturally into two regions: an inner region near the body or the airfoil, and an outer region near the walls. From the inner region the wall is far away; its effect is small, so that the flow is sonic free flow plus a small correction. It is this correction we want to determine that represents the wall interference. The outer region must account for the choked nature of the flow, namely the sonic line meeting the wall. We construct asymptotic expansions in each of these regions, then match to find (1) the relationship between H and K_T to maintain choking, (2) the order of magnitude of the correction to sonic free flow near the body, and (3) an explicit boundary value problem to be solved for this correction. The result is, for 2D flows, $K_T = O(H^{-2/5})$, and the wind-tunnel wall correction is $O(K_T^3)$; for slender body flows $K_T = O(H^{-6/7})$, and the correction is $O(K_T^{5/3})$.

Here the results are only summarized. Underlined terms are those determined by matching . For details see [2], [3]. Note that all analysis is carried out in the physical plane. We discuss first the case of 2D flow.

Inner region: $(x, \tilde{y})$ fixed, $K_T \to 0, H \to \infty$.

$$(4.10) \qquad (\gamma+1)\phi(x,\tilde{y};K_T,H) = \phi_0(x,\tilde{y}) + K_T x + \underline{K_T^3 \phi_1(x,\tilde{y})} + \cdots$$

where ϕ_0 is sonic flow so that (in front of the shock)

$$(4.11) \qquad -\phi_{0x}\phi_{0xx} + \phi_{0\tilde{y}\tilde{y}} = 0,$$

$$(4.12) \qquad \phi_{0\tilde{y}}(x, 0\pm) = F'_{u,\ell}(x) \quad \text{for} \quad 0 \le x \le 1,$$

$$(4.13) \qquad \phi_{0x} \to 0 \quad \text{as} \quad x \to -\infty\,.$$

In fact

$$\phi_0 \sim \tilde{y}^{2/5}\frac{f(a\xi)}{a^3} + c_0 + c_1\tilde{y}^{-1/5}\frac{f_1(a\xi)}{a^3} + \cdots \quad \text{as} \quad |\tilde{y}| \to \infty\,,$$

and the boundary value problem for ϕ_1 is

$$(4.14) \qquad -(\phi_{0x}\phi_{1x})_x + \phi_{1\tilde{y}\tilde{y}} = 0,$$

$$(4.15) \qquad \phi_{1\tilde{y}}(x, 0\pm) = 0 \quad \text{for} \quad 0 \le x \le 1,$$

$$(4.16) \qquad \phi_1 \sim \underline{C_1\frac{\tilde{y}^{8/5}F(a\xi)}{a^3}} \quad \text{as} \quad |\tilde{y}| \to \infty\,.$$

Here $\xi = x/\tilde{y}^{4/5}$, f, F, and f_1, which will be needed later, are known and most easily given in parametric form

(4.17)

$$a\xi = \frac{1}{2}a_1^{1/5}s^{-2/5}(2s-1), \quad a_1 = 2^9 3^3 5^{-5}, \quad 0 \le s \le \frac{1}{6}(5\sqrt{3}+8),$$

$$f = \frac{1}{6}a_1^{3/5}s^{-1/5}(2s^2-3s+6), \quad f_1 = (\sqrt{3}/2)s^{3/5}, \quad F = s^{-4/5}(1-6s).$$

The constants a and C_1 depend on the shape of the body and can be determined, for a specific body, by conservation laws [4], [8].

Outer region: $\bar{x} = K_T^2 x$, $\bar{y} = K_T^{5/2}\tilde{y}$ fixed as $K_T \to 0$, $H \to \infty$, and we obtain $\phi = (1/K_T)\bar{\phi}(\bar{x}, \bar{y}; K_T)$. Note that $\tilde{y}$ is stretched so that we stay near the wall, and the other stretchings follow to maintain the nonlinear problem and similarity variables in the near field. The boundary value problem for $\bar{\phi}$ is

$$(1-(\gamma+1)\bar{\phi}_{\bar{x}})\bar{\phi}_{\bar{x}\bar{x}} + \bar{\phi}_{\bar{y}\bar{y}} = 0, \tag{4.18}$$

$$\bar{\phi}_{\bar{y}}(x, \pm\bar{H}) = 0, \tag{4.19}$$

$$\bar{\phi}_{\bar{x}} \to 0 \quad \text{as} \quad \bar{x} \to -\infty, \tag{4.20}$$

with $\bar{H} = K_T^{5/2}H$. The outer expansion is

$$(\gamma+1)\bar{\phi}(\bar{x}, \bar{y}; K_T) = \psi_0(\bar{x}, \bar{y}) + \cdots \tag{4.21}$$

and ψ_0 satisfies the boundary value problem

$$(1-\psi_{0\bar{x}})\psi_{0\bar{x}\bar{x}} + \psi_{0\bar{y}\bar{y}} = 0, \tag{4.22}$$

$$\psi_{0\bar{y}}(x, \pm\bar{H}) = 0, \tag{4.23}$$

$$\psi_{0\bar{x}} \to 0 \quad \text{as} \quad \bar{x} \to -\infty\,, \tag{4.24}$$

and the condition that the sonic line reaches the wall. Using matching conditions and (4.22)–(4.24) it follows that

$$\psi_0 \sim \underline{\bar{y}^{2/5}\frac{f(a\xi)}{a^3}} + \bar{x} + C_1\bar{y}^{8/5}\frac{F(a\xi)}{a^3} + \cdots \quad \text{as} \quad |\bar{y}| \to 0\,. \tag{4.25}$$

So, since the linear term does not contribute to lift, from (4.10) the correction to sonic flow near the body is $O(K_T^3)$. In order to find it explicitly we must solve (4.14)–(4.16) for ϕ_1. However, to do that we must know C_1, which is determined from the outer problem (4.22)–(4.24) which also determines $\bar{H}$.

$\bar{H}$ and C_1 can be determined analytically [2] by conservation laws. In fact

$$\bar{H} = 2^{-5/3}3^{7/6}5\mathcal{K}\pi\,, \quad a^{-5} = \frac{5^5}{2^{23/3}3^{4/3}}\,\mathcal{K},$$

and

$$C_1 = \hat{C}_1\bar{H}^{-6/5}$$

where

$$\hat{C}_1 = 2^{148/15}2^{-55}3^{23/30}5^{-4}11^{3/2}(0.0260)$$

and $\mathcal{K}$ (or a) is determined by the specific body shape [4], [8]. Hence now to find the explicit wind-tunnel correction we must solve (numerically) the linear boundary value problem for ϕ_1 which requires knowledge of ϕ_0, the solution to the nonlinear problem. This is in progress.

The analysis for the 3D case is similar, but with one crucial difference. In that case we cannot yet find the analogues of $\bar{H}$ and C_1 analytically since the wealth of conservation laws for 2D is not available for axisymmetric flow. Thus at this point the analogue of C_1 must be found numerically. As was noted above the wind-tunnel wall effect is much stronger in 3D flows namely $O(K_T^{5/3})$ as opposed to $O(K_T^{\frac{3}{2}})$.

4.4. Relationship to Nonchoked Flow

Transonic solid wall wind-tunnel interference has been examined by, among others, Malmuth et al. [6], [11], for slow (transonic) flow, that is, for Mach numbers not near choking. In this case it is found that the correction to free flight 2D flow due to the walls is $O\left(1/H^2\right)$. Clearly the transonic correction $O\left(1/H^{6/5}\right)$ is a much stronger effect. For slender bodies (3D) flows in axisymmetric wind-tunnels the correction was found to be $O(H^{-3})$.

4.5. Previous Work and Future Work

Since the earlier analyses (namely those of Guderley, Marschner, Sevost'ianov, and Morioka) were for 2D flows and took place not in the physical plane, as in the last sections, but rather in the hodograph plane, we first introduce the hodograph problem. We do so for the small disturbance formulation, but the analysis (using the Legendre transform of the potential) is virtually identical for the full potential formulation. We also do the symmetric, double wedge since all the above analyses were for symmetric flow and all but one for the double wedge. The advantage of the latter is that the image of the airfoil is known in the hodograph plane a priori, and is in fact very simple.

In the hodograph plane with

$$w = (\gamma+1)\phi_x - K_T, \quad \theta = (\gamma+1)\phi_{\tilde{y}}\,,$$

$$\left\{\begin{array}{lcll} ww_x & - & \theta_{\tilde{y}} & = 0 \\ w_{\tilde{y}} & - & \theta_x & = 0 \end{array}\right\} \text{ becomes } \left\{\begin{array}{lcll} w\tilde{y}_\theta & - & x_w & = 0 \\ \tilde{y}_w & - & x_\theta & = 0 \end{array}\right\}$$

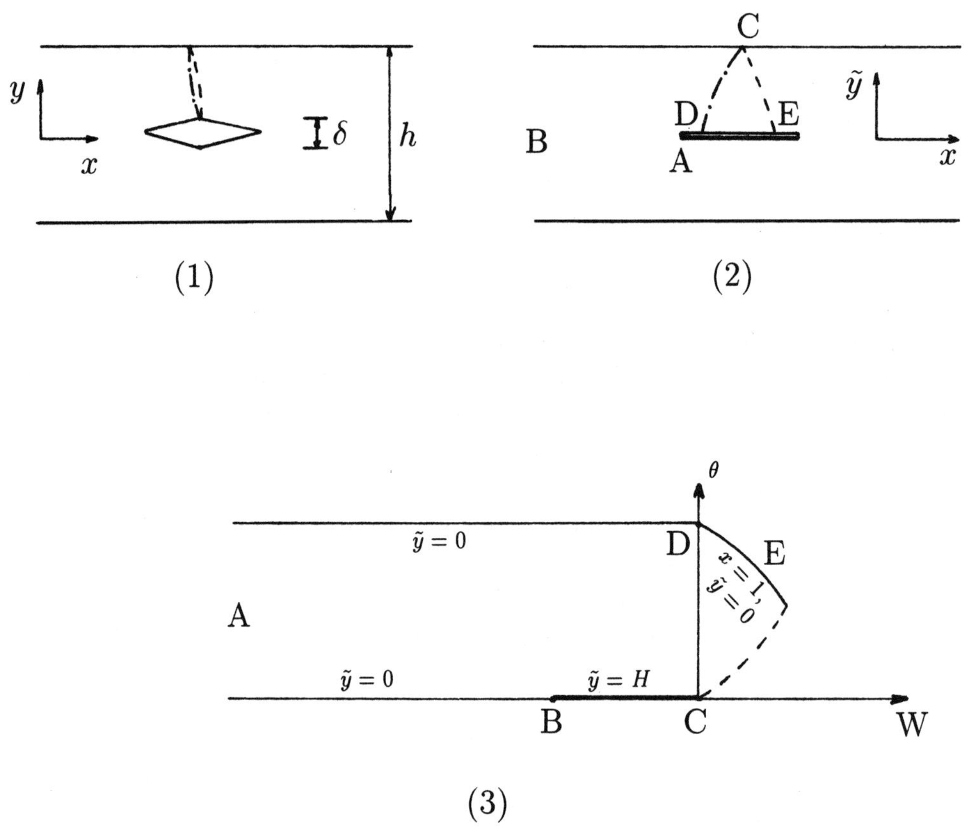

FIG. 4.3. *Symmetric double wedge wind-tunnel flow in* (1) *original physical plane,* (2) *small disturbance plane,* (3) *hodograph plane.*

where

$$\left\{ \begin{array}{lcl} x_\theta & = & -\frac{1}{j} w_{\tilde{y}} \\ x_w & = & \frac{1}{j} \theta_{\tilde{y}} \end{array} \right\} \text{ and } \left\{ \begin{array}{lcl} \tilde{y}_\theta & = & \frac{1}{j} w_x \\ \tilde{y}_w & = & -\frac{1}{j} \theta_x \end{array} \right\},$$

or

$$w\tilde{y}_{\theta\theta} - \tilde{y}_{ww} = 0 \tag{4.26}$$

and $j = \partial(w,\theta)/\partial(x,\tilde{y}) = w_x\theta_{\tilde{y}} - \theta_x w_{\tilde{y}}$, $J = 1/j = x_w\tilde{y}_\theta - x_\theta\tilde{y}_w$. Of course the mapping is only valid for $j \neq 0, \infty$ which will be discussed below.

Since the flow is symmetric, $\phi_{\tilde{y}} = 0$ along $\tilde{y} = 0$, and we need only solve for the flow in $\tilde{y} \geq 0$. Thus $\tilde{y} = 0$, $-\infty < x < 0$ maps in the hodograph to $\theta = 0$, $-\infty < w < -K_T$, and the front upper half of the body to $\theta = \theta_0 = \gamma + 1$ for $-\infty < w < 0$. On the wall $\phi_{\tilde{y}} = 0$, hence the wall maps to $\theta = 0, -K_T < w < 0$. The sonic line and limit characteristic emanate from the wedge shoulder so that the expansion DE maps to the corresponding characteristic in the hodograph. The problem to solve in the hodograph plane is then Tricomi's equation (4.26) subject to

$$\tilde{y} = 0 \quad \text{on} \quad \theta = 0, \quad -\infty < w < -K_T, \tag{4.27}$$

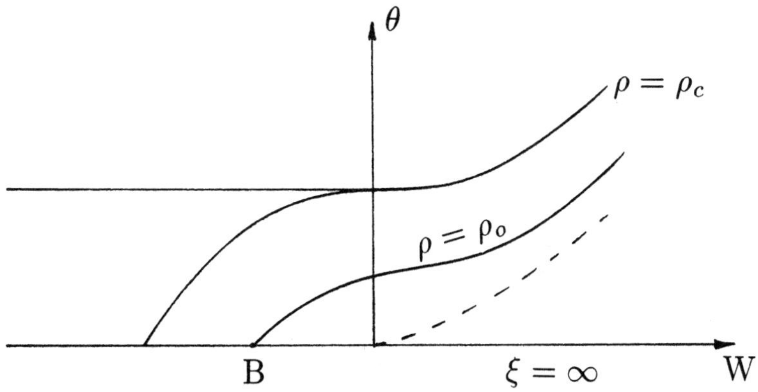

FIG. 4.4. $\xi = w/(\frac{3}{2}\theta)^{2/3}$; $\rho = -w^3 + \left(\frac{3}{2}\theta\right)^2$.

(4.28) $$\tilde{y} = 0 \quad \text{on} \quad \theta = \gamma + 1, \quad -\infty < w < 0,$$

(4.29) $$\tilde{y} = H \quad \text{on} \quad \theta = 0, \quad -K_T < w < 0,$$

(4.30) $$\tilde{y} = 0 \quad \text{on DE, namely on} \quad \theta + 2/3w^{3/2} = \theta_0 = \gamma + 1,$$

i.e.,

(4.31) $$\text{on} \quad 0 < w < (3/4(\gamma + 1))^{2/3}$$

with the appropriate singularity at B. The problem as it stands is well posed but we have the additional restriction that $x = 1$ on DE. Thus the correct parameter combination K_T, H must be found so that the solution of (4.26)–(4.31) in fact corresponds to $x = 1$ at D, that is,

(4.32) $$\int_{-\infty}^{0} x_w \, dw = \int_{-\infty}^{0} w\tilde{y}_\theta \, dw = 1.$$

Note that the form of the singularity at B can be found explicitly.

The first serious attempts to deal with choked flow wind-tunnel wall interference appear to be those of Marschner [12], which were motivated by earlier work of Guderley. Marschner considered 2D flow about a nonlifting (symmetric) double wedge airfoil in a wide solid wall wind-tunnel at choking and compared his results to those for a double wedge in a sonic free stream or in a sonic free jet. As pointed out above the advantage of studying 2D flow is that the hodograph formulation can be used, and the advantage of studying a double wedge airfoil is that the image of the boundary in the hodograph plane is known a priori. Thus Marschner studied the problem in the hodograph (θ, w)-plane (in his case θ, η). He constructed eigenfunction expansions of the solutions in terms of the transformed coordinates $\xi = w/(3\theta/2)^{2/3}$, $\rho = -w^3 + (3\theta/2)^2$ which allow for separation of variables of the equation.

In these coordinates Tricomi's equation for $\tilde{y}$ becomes

$$\frac{(1-\xi^3)^2}{\xi}\tilde{y}_{\xi\xi} - \frac{5}{2}\xi(1-\xi^3)\tilde{y}_\xi + \frac{\tilde{y}}{16} = 9\rho^2\tilde{y}_{\rho\rho} + \frac{21}{2}\rho\tilde{y}_\rho + \frac{1}{16}\tilde{y}$$

so that product solutions exist of the form

$$\tilde{y} = \rho^{-\frac{1}{12}\pm\sqrt{\frac{\lambda}{3}}}\, G(\xi)$$

where

$$\frac{d}{d\xi}\left((1-\xi^3)^{5/6}\frac{dG}{d\xi}\right) + \left(\frac{1}{16} - \lambda\right)\frac{\xi}{(1-\xi^3)^{7/6}}G = 0\,.$$

Under appropriate boundary conditions on G the eigenfunctions will form a complete set. Series expansions were formulated separately in an inner and an outer region which were clearly divided by a specified curve $\rho = \rho_0$, namely the $\rho =$ constant curve passing through B. Instead of asymptotic expansions or asymptotic matching, coefficients in the infinite series were found numerically by equating the value of the series and its normal derivative along $\rho = \rho_0$.

There are some difficulties and open questions with the procedures and analysis. However, the results do indicate that $K_T \propto H^{-2/5}$ and that the correction to sonic free flow is of order $H^{-6/5}$ as we found. Actual computations are shown for $\delta = .1, H = 100/13$.

The next go at the problem was by Morioka [13] who also studied the symmetric, nonlifting double wedge airfoil in a solid wall wind-tunnel. Morioka calculated the entire flow numerically using Vincenti's relaxation method in the hodograph. His results are computed for $\delta = .1$ and three particular Mach numbers, $M_\infty = .869, .829, .795$. The results seem to agree with those of Marschner.

Then in 1965 Sevost'ianov considered a symmetric, nonlifting, convex, 2D airfoil in choked wind-tunnel flow. Again the transonic small disturbance equation was analyzed in the hodograph plane. In this paper Sevost'ianov noticed that for a wide channel the solution would be only weakly affected by the body shape. He then obtained the streamfunction representation valid far from the body. His method of solution was using singular integral equations. Basically Sevost'ianov writes $\tilde{y} = Y + y_3$. Here Y satisfies the hodograph problem (4.26)–(4.31) and in place of $\tilde{y} = 0$ on the body profile he puts $Y \to 0$ at infinity. Then y_3 must be found to not alter the value of Y along $\theta = 0$, but to correct Y so that $\tilde{y} = 0$ on the image of the profile. He does not find y_3 but does find Y and shows that as $H \to \infty$, if choking is maintained, Y approaches Y_0, the solution for 2D sonic free flow. Thus he writes

$$Y(\theta,\eta) = y_1(\theta,\eta) + y_2(\theta,\eta),$$

$$y_1(\theta,\eta) = Ht_\infty^{2/3}t^{1/3}\int_0^\infty e^{-\lambda\theta}\,\mathrm{J}_{-\frac{1}{3}}(\lambda t)\,\mathrm{J}_{2/3}(\lambda t_\infty)\,d\lambda,$$

$$y_2(\theta,\eta) = -\tau_0\int_0^\infty \nu(s)\,\{[(\theta-s)^2+\frac{4}{9}\eta^3]^{-1/6} - [(\theta+s)^2+\frac{4}{9}\eta^3]^{-1/6}\}\,ds$$

where τ_0 is a constant,

$$\nu(\theta) = \theta^{-1/3} \frac{Ht_\infty^{5/3}}{t_\infty^2 + \theta^2} \frac{4(2)^{1/3}}{5\pi(3)^{1/6}} F\left(\frac{1}{6}, 1; \frac{11}{6}; \frac{t_\infty^2}{t_\infty^2 + \theta^2}\right),$$

$t = 2\eta^{3/2}/3$, $t_\infty = 2K_T^{3/2}/3$ and

$$Y_0(\theta, \eta) = C\rho^{-5/3}\{(1-s)^{1/3}\left(\frac{1}{3}+s\right) - (1+s)^{1/3}\left(\frac{1}{3}-s\right)\}, \quad s = \frac{\theta}{\rho},$$

where $\rho = \sqrt{\theta^2 + 4\eta^3/9}$ and C is a constant related to the size of the wing.

In particular he finds in the limiting process that

$$H = O(1 - M_\infty)^{-5/2},$$

which agrees with previous results.

Recently Schleiniger considered the actual numerical dependence of H on K_T for the case of a double-wedge, nonlifting airfoil (2D). The hodograph formulation was used and the solution decomposed as the sum of a singular part and a regular part. The singular part of the solution satisfying the boundary conditions on the subsonic part ($w < 0$) can be found explicitly and, for $w > 0$, after analytical continuation it is

$$\tilde{y}_s(w, \theta) = H\tau^{1/3}\tau_\infty^{2/3} \int_0^\infty \frac{\sinh \lambda(\theta_0 - \theta)}{\sinh \lambda\theta_0} \mathrm{J}_{2/3}(\lambda\tau_\infty)\,\mathrm{I}_{-1/3}(\lambda\tau)\, d\lambda$$

where $\tau = (2/3)w^{3/2}$ and $\tau_\infty = (2/3)K_T^{3/2}$. Note that it is a modification of $y_1(\theta, \eta)$ given by Sevost'ianov.

The regular part of the solution is found by eigenfunction expansions with coefficients determined through the condition (4.30) on the characteristic boundary. Then (4.32) is enforced, which gives the desired dependence $H = \mathcal{F}(\tau_\infty)$.

As a first approach, a collocation method along the characteristic boundary was used to calculate the coefficients in the eigenfunction expansion. But the convergence is very slow and the numerical results are not sufficiently accurate. A new approach, following ideas similar to those of Guderley and Yoshihara [10], is currently under implementation. They calculated the flow profile over a double-wedge nonlifting airfoil at Mach number one in an infinite plane. Here, for wind-tunnel flow, similar ideas are used. The solution is extended, in the hodograph plane, to include the supersonic "triangle" made up by the characteristic boundary ($0 < w < (3\theta_0/4)^{2/3}$, $\theta = \theta_0 - 2w^{3/2}/3$), the horizontal segment ($\theta = \theta_0$, $0 < w < (3\theta_0/4)^{2/3}$), and the vertical segment ($w = (3\theta_0/4)^{2/3}$, $\theta_0/2 < \theta < \theta_0$). The boundary conditions for this extended domain are $\tilde{y} = 0$ on both the vertical and horizontal segments described above, and $\tilde{y}_w = 0$ on the vertical segment. This will actually result in a solution that is identically equal to zero in the triangle in question.

Finally, to justify working in the hodograph plane it must be shown that $J \neq 0$ so that the mapping back to the physical plane holds globally. This particular problem has not been attacked previously but, in doing the numerical calculation, the value of J can be monitored to determine, at least to within the accuracy of the calculation, whether there are limit lines.

References

[1] S. B. Berndt, *Theory of wall interference in transonic wind-tunnels*, pp. 288–309.
[2] J. D. Cole and L. P. Cook, *Two-dimensional choked transonic flow*, Z. Angew. Math. Phys., 39 (1988), pp. 1–12.
[3] ——, *Transonic choking and stabilization for flows about slender bodies*, Mech. Res. Comm., 15(4) (1988), pp. 213–219.
[4] ——, *Transonic Aerodynamics*, North-Holland, Amsterdam, 1986.
[5] J. D. Cole and N. D. Malmuth, *Shock wave location on a slender transonic body of revolution*, Mech. Res. Comm., 16(6) (1989), pp. 353–358.
[6] J. D. Cole, N. D. Malmuth, and F. J. Zeigler, *An asymptotic theory of solid tunnel wall interference on transonic airfoils*, AIAA/ASME Joint Thermophysics, Fluids, Plasma and Heat Transfer Conference, St. Louis, MO, June 7–11, 1982.
[7] L. P. Cook and F. J. Zeigler, *The stabilization law for transonic flows*, SIAM J. Appl. Math., 46 (1986), pp. 27–48.
[8] P. Germain, *Écoulements transsoniques homogènes*, tech. paper 242, ONERA, 1965.
[9] K. G. Guderley, *The Theory of Transonic Flow*, Pergamon Press, Oxford, 1962.
[10] K. G. Guderley and H. Yoshihara, *The flow over a wedge profile at Mach number 1*, J. Aero. Sci., Nov. (1950), pp. 723–735.
[11] N. D. Malmuth and J. D. Cole, *Study of an asymptotic theory of transonic wind-tunnel wall interference*, tech. report AEDC-TR-84-8, AEDC, 1984.
[12] B. W. Marschner, *The flow over a body in a choked wind-tunnel and in a sonic free jet*, J. Aero. Sci., April (1956), pp. 368–376.
[13] S. Morioka, *High subsonic flow past a wedge in a two-dimensional wind-tunnel at its choked state*, J. Phys. Soc. Japan, 14(8) (1959), pp. 1098–1101.
[14] G. D. Sevost'ianov, *Two-dimensional transonic gas flow far from a profile located in the channel*, Prikl. Mat. Mekh., 29(5), pp. 863–869, Transl. (1965), pp. 1021–1028.

Chapter 5

Some Applications of Combined Asymptotics and Numerics in Fluid Mechanics and Aerodynamics

N. D. Malmuth†

5.1. Introduction

In the 1950s at Caltech and elsewhere, there was a resurgence in interest in perturbation methods, particularly those oriented to singular perturbation problems. As described in [1]–[3] successful applications of the technique have been numerous. With the revolution in high speed computers, attention has been until recently diverted away from these approaches. A conviction still held in many circles is that asymptotics are of low relative utility compared to computational procedures.

Strong evidence exists that there is much to be gained from an interplay between both asymptotic and numerical viewpoints. On one hand, the asymptotics illuminate the essential scales and similitudes of the problem. They also provide insights into invariances and physics that are not transparent from a purely numerical solution. Moreover, asymptotic methods give us an idea about parametric dependencies, bring out the essential features of the problem, and simplify the formulation (even reducing its dimensionality), as well as providing information about the smooth, consistent blending of temporal and spatial regions where different effects are important. In fact, success of this matching process often serves as a check of the correctness of the asymptotic procedure itself.

On the other hand, modern computational techniques yield a powerful means of dealing with nonlinearities intractible to analytical methods. Even here, however, new insights and efficient numerical approaches can be derived by exploiting asymptotic ideas. Some of this is related to older numerical techniques such as "subtracting or factoring out" the singularity that can be rep-

† A portion of the efforts described here were supported by NASA Contract NASA9–14000 as well as Air Force Contracts F40600–84–C–0010 and F46920–92–C–0006DEF.

resented in many instances as a coordinate rather than parameter expansion. Other ideas currently receiving attention are use of asymptotic behaviors in discretization approximations for derivatives, domain decomposition, and convergence acceleration techniques using solution initialization and defect minimization with asymptotics. Even gridding issues associated with numerical solutions can be resolved by asymptotic methods. An example is the viscous boundary layer of high Reynolds number flow whose length scale is determined from asymptotics and similarity. These suggest the grid clustering necessary to resolve the fine structure of the boundary layer as its thickness and the viscosity become small.

This article will survey examples that illustrate the effectiveness of combined asymptotics and numerics (CAN) methods. The view is primarily the one first mentioned here, namely, allowing the asymptotics to provide an economical formulation that is treated by the numerics. The second viewpoint involving somewhat of a more intimate relationship, coined by some authors as "asymptotically induced numerics" will not receive as much emphasis in what follows. It, however, is a very active field of research.

Examples to be discussed come from transonic flow and an application of lubrication theory appropriate to foil bearings, paper making, as well as the production of film and magnetic recording. All of these applications involve thin layers. Some of them contain diverse length scales. The asymptotics are called "limit-process" asymptotics in the sense that the expansions are increasingly accurate approximations of the solution in a "distinguished limit" as a parameter tends to a limit. Cole and Kevorkian [4] provide formal definitions of these representations, elucidate the basic theoretical ideas, and discuss various applications. All the numerical methods illustrated are of inconsequential to very low computational intensity, illustrating the advantage of asymptotic "preprocessing."

5.2. Transonic Slender Body Theory and the Area Rule

In transonic flow, a limiting process can be defined in terms of the free stream Mach number M_∞ and the body thickness ratio δ that measures the slenderness of the body. Useful theories based on this limit can be developed for practically important low-wave drag bodies exemplified by missiles and fighters such as the Advanced Tactical Fighter (ATF). Other slender vehicles such as the hypersonic National Aerospace Plane (NASP) need to accelerate through the speed of sound to achieve orbital velocity V_O. Drag at the "transonic pinch point" critically affects attainment of V_O, since accelerations controlled by drag peaks will affect influential parameters such as fuel consumed, payload, structural and vehicle takeoff weight. Quick response methods are needed to help reach V_O as well as to connect the vehicle shape with its aerodynamic performance. Although computational fluid dynamics (CFD) treats this nonlinear regime, CAN methods offer a lower computational intensity alternative for preliminary aerodynamic design and vehicle sizing, providing quick evaluation of mission and configurational impacts of hundreds of parametric changes and accurate modeling of the nonlinear physics.

Oswatitsch [5] and others developed the Transonic Equivalence Rule (TER) in which a complicated shape could be replaced by an equivalent body of revolution under mild constraints for purposes of defining the flow field over the more complicated body. This rule was clarified in [6] by a demonstration of how it is embedded in an approximation procedure based on the aforementioned distinguished limit leading to a model called Transonic Slender Body Theory (TSBT). Great practical impacts of a TER adjunct called the Transonic Area Rule (TAR) were developed in the pioneering work of Whitcomb [7] and Jones [8].

Figure 5.1 shows a body of thickness ratio δ in a cylindrical polar coordinate system appropriate to limit processes giving the TER. For this system, the flow consists of three regions of different character. The precise asymptotic limits that dilineate these zones are given in [6], where it is shown that there is an "inner" region near the body where crossflow gradients dominate. In the cylindrical coordinate system shown in Fig. 5.1, if r and x denote dimensionless radial and axial coordinates respectively, an inner limit is defined in which the crossflow dominated region appears invariantly as $M_\infty \to 1$. As in TSBT, this limiting behavior is expressed in terms of a transonic similarity parameter K. The inner limit is

$$K = \frac{1 - M_\infty^2}{\delta^2}, \quad r^* = \frac{r}{\delta} \text{ fixed as } \delta \to 0. \tag{5.1}$$

If Φ is the velocity potential, an inner expansion for Φ in the limit process (5.1) gives harmonic behavior in the crossflow plane for a perturbation potential ϕ, where the expansion is

$$\Phi(x, r, \theta; M_\infty, \delta) = U(x + 2\delta^2 \log \delta \; S^*(x) + \delta^2 \phi_2^*(x, r^*; K) + O(\delta^4)). \tag{5.2}$$

In (5.2), U is the freestream speed and $S^*(x)$ is a source distribution to be determined by matching.

Away from the body of Fig. 5.1, asymmetric effects decay and nonlinear phenomena become important. This occurs on a lateral scale of the steeply inclined Mach waves in supersonic flow zones. This region is defined by the "outer" limit

$$K = \frac{1 - M_\infty^2}{\delta^2}, \quad \tilde{r} = r\delta \text{ fixed as } \delta \to 0. \tag{5.3}$$

The appropriate asymptotic behavior for Φ in the outer region is

$$\Phi(x, r, \theta; M_\infty, \delta) = U(x + \delta^2 \phi(x, \tilde{r}; K) + O(\delta^4)). \tag{5.4}$$

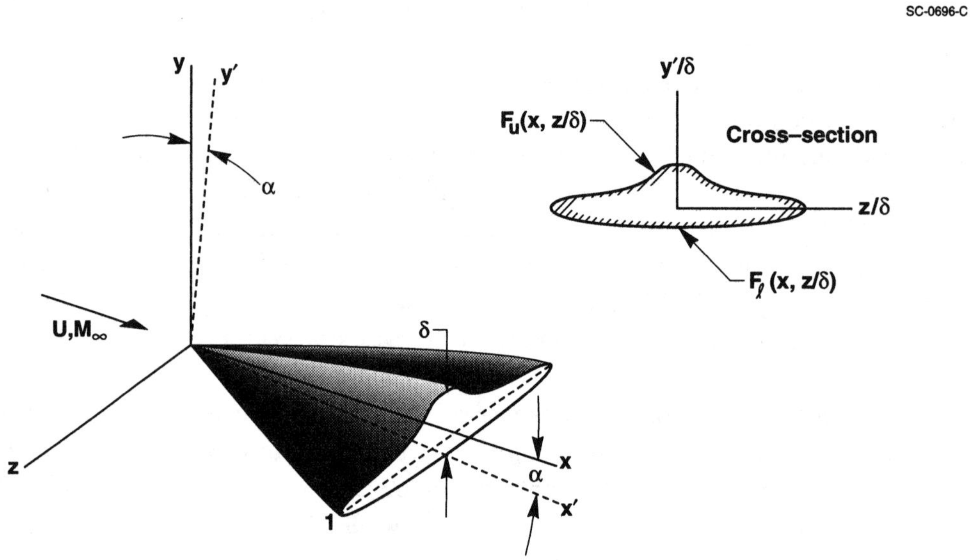

FIG. 5.1. *Transonic body.*

Expansions (5.2) and (5.4) only match to dominant orders. Cole and Cook [6] show the need for an intermediate expansion to obtain higher order refinements. Figure 5.2 shows how the inner and outer expansions are combined according to matching to give a uniformly valid expression for the pressure distribution over an asymmetric body. The nonlinear harmonic boundary value problem for ϕ_2^* is shown in the near field (inner) zone for the body given by

(5.5) $$r = \delta F(x, \theta),$$

In the coordinates defined in the left-hand side of the figure, c is reference length such as the wing chord. The pressure coefficient C_P given by (F1) in the figure is partially determined in terms of ϕ_{2x}^* and the squared resultant crossflow speed ${v_B^*}^2 + {w_B^*}^2$ which is related to gradients of ϕ_2^* in (F2).

The dominant term in (F1) is the second streamwise derivative of the cross sectional area $A(x)$. The last remaining component of the pressure $2g'(x)$in (F1) comes from the nonlinear axisymmetric outer potential ϕ that is governed by the axisymmetrical Karman transonic small disturbance equation (F3) subject to the boundary condition (F4) which is one of the essential results of dominant order matching. Another is the formula for $g(x; K)$, (F5).

These relations are an excellent illustration of the simplifications afforded by the TER, which reduces the 3D flow problem to an axisymmetric one. By integration, it is evident that the wave drag only depends on the area distribution of the equivalent body of revolution. This is the transonic area rule. By this reduction in dimensionality, the original computational problem is sizably reduced to the numerical treatment of the axisymmetric small disturbance equation (F3) and a boundary integral method for the near field harmonic

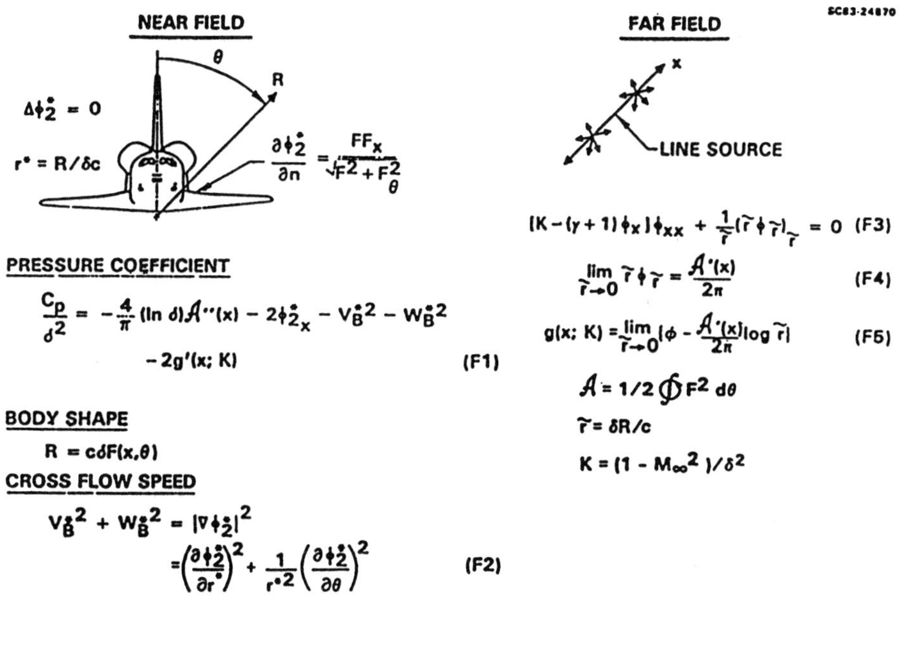

FIG. 5.2. *Transonic equivalence rule.*

boundary value problem. A 2D "panel" approach that treats each x station body contour by a sequence of straight line sources around the body contour works quite well and is a practically useful method. Modern techniques can increase the speed and accuracy of inverting the full matrix discretization of the panel method singular integral equation. If N is the approximating number of segments, the operation count can be reduced from $O(N^2)$ to $O(N \log N)$ using these methods.

As an illustration of the power of this approach, Fig. 5.3 shows a comparison reported in [9] of pressures over a wing body combination as predicted by the TER with experiment. Figures 5.4 and 5.5 show comparisons at slightly supersonic Mach numbers for the Shuttle Orbiter. Figure 5.5 gives pressures on the top centerline of the Orbiter fuselage at M_∞=1.05. These results suggest encouraging elasticity of the theory for not-so-slender shapes. This is another practical bonus of CAN approaches when the underlying asymptotics have such favorable properties. Another application of the TER and TAR is shown in Fig. 5.6, for the wave drag rise of an advanced fighter configuration. Under certain constraints indicated in [6], the TAR for a body is

$$\frac{D}{\rho U^2} = \int_0^1 A''(x) g(x)\, dx, \tag{5.6}$$

where D and ρ are the drag and density, respectively. The figure shows the ability of the CAN TAR to give good correlations with experiments that have been not been reported to our knowledge with pure CFD approaches. Appropriate filtering techniques are needed for the numerical evaluation of $A''(x)$.

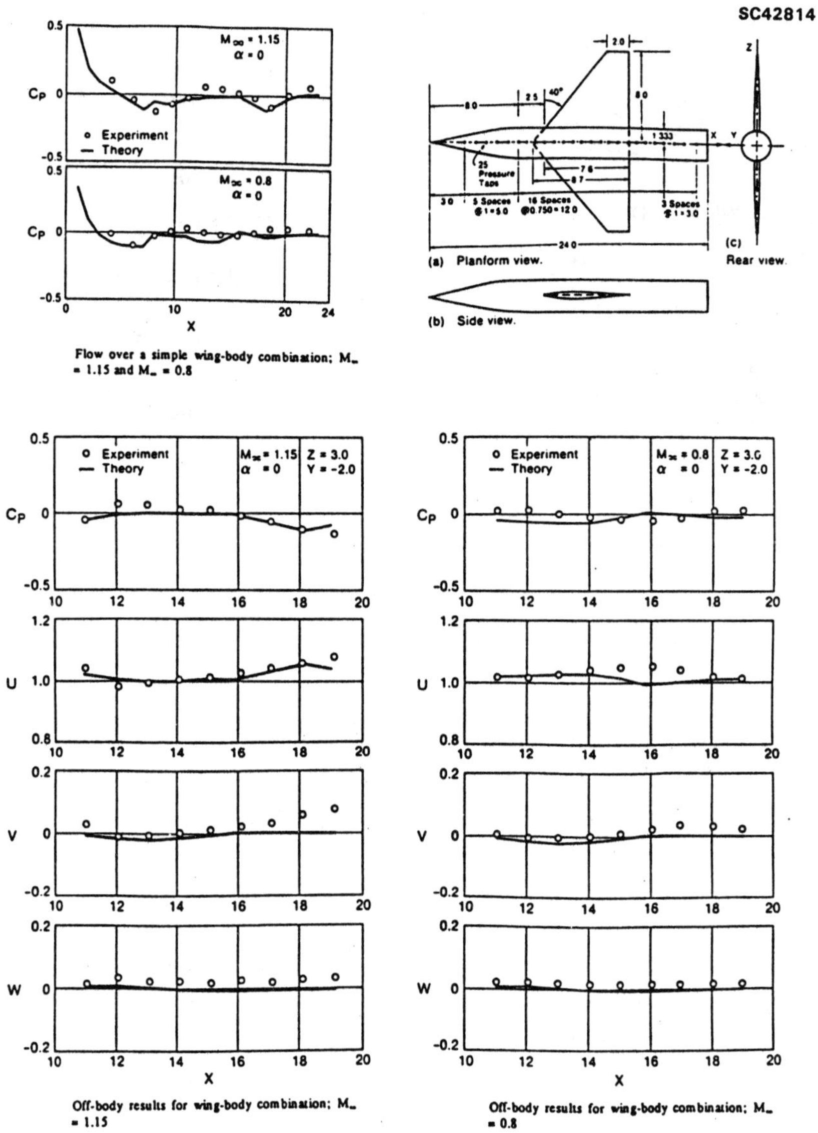

FIG. 5.3. *Comparison of TER with experiment for generic wing body,* $M_\infty = .8$ *and* 1.15.

International Mathematical Software Library (IMSL) spline smoothing routines provide effective tools for this purpose.

5.3. Shocks over Transonic Bodies of Revolution

An interesting illustration of the power of asymptotics to give surprising insight into the physics of the flow is application of the slender body theory just discussed (TSBT) to prediction of the position of shocks over bodies of

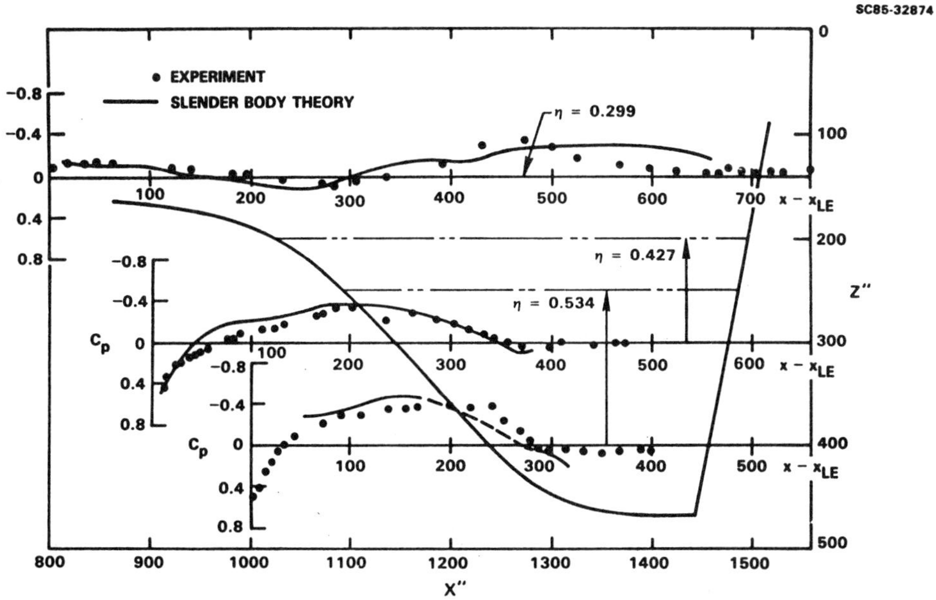

FIG. 5.4. *Comparison of TER with experiment for Shuttle Orbiter Wing at* $M_\infty = 1.05$.

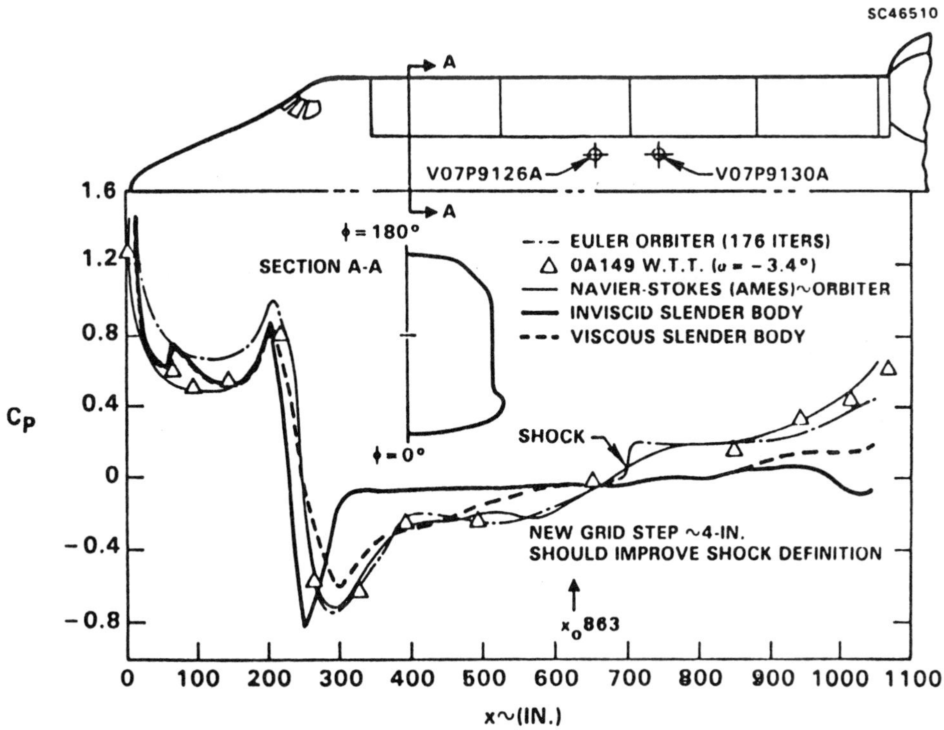

FIG. 5.5. *Comparison of TER with experiment for Shuttle Orbiter fuselage.*

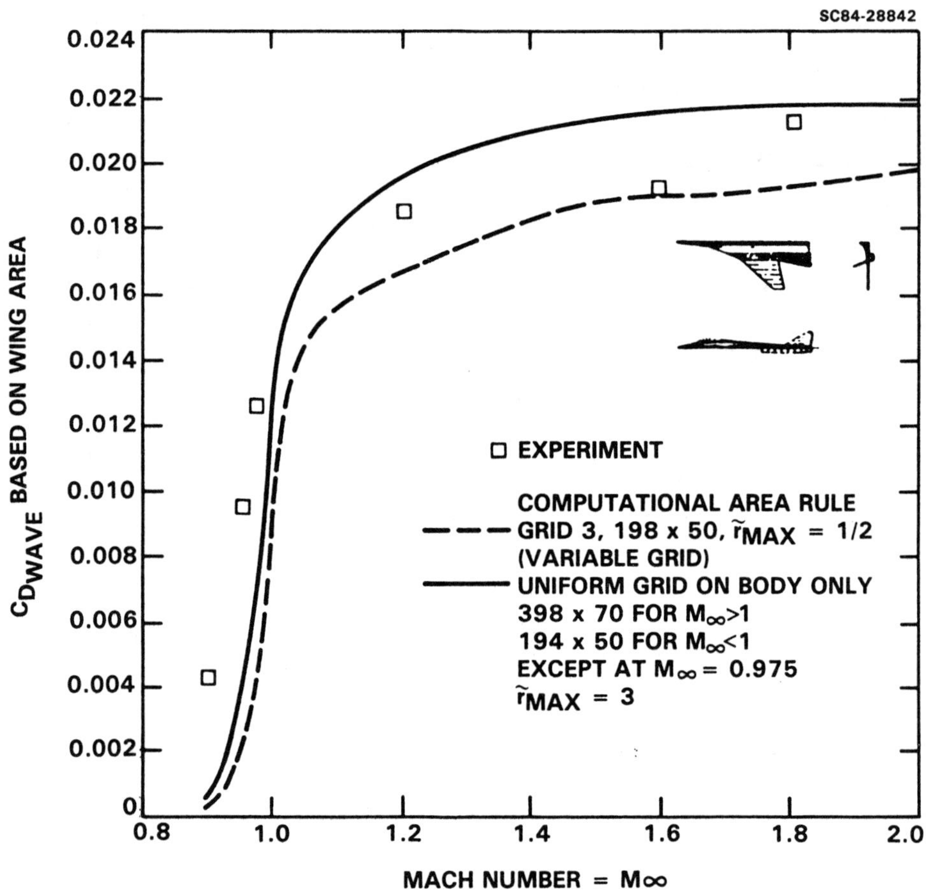

FIG. 5.6. *Comparison of TAR predicted wave drag rise with experiment.*

revolution. Here, in contrast to planar flows, the theory indicates that the shock position of almost sonic axisymmetric bodies depends only on the properties of the streamwise area progression and occurs at the axial location where there is a point of inflection in that distribution. This property is seen from a simple-minded argument coming from unpublished studies of Malmuth and C. C. Wu. A more detailed analysis is given in [10], which has recently been further refined by J. D. Cole.

In the simple-minded argument, if ϕ is the outer perturbation potential, $A(x)$ the cross-sectional area, and x the streamwise Cartesian coordinate, the inner expansion

$$\phi \doteq \frac{A'(x)}{2\pi}\log r + g(x) + \cdots \quad \text{as } r \to 0 \tag{5.7}$$

holds as well as the Prandtl relation at the foot of the almost normal shock near its intersection with the body,

$$\langle \phi_x \rangle = \frac{K}{\gamma + 1} \tag{5.8}$$

$$(K = \frac{1 - M_\infty^2}{\delta^2} = \text{Transonic Similarity Parameter})$$

$$(\langle\ \rangle \doteq \text{average across shock }),$$

where M_∞ = freestream Mach number, δ = body thickness ratio.

Differentiating (5.7) with respect to x and forming the shock average gives

$$\langle \phi_x \rangle \doteq \frac{A''(x)}{2\pi} \log r + \langle g(x) \rangle + \cdots \quad \text{as } r \to 0. \tag{5.7$'$}$$

For order of magnitude consistency, (5.8), which is applicable to near unity Mach numbers, implies

$$A''(x) = 0 \tag{5.9}$$

in (5.7′) at the shock's foot. *This is the aforementioned shock position rule.* Shock locations predicted from (5.9) have been compared with data from [11] for the bodies therein. Consistent with the notation of [11] in which $\xi = \frac{x}{l}$, l = body length, denote this location as $\xi_{A''=0}$, and if the location ξ_{MAX} of the maximum thickness is such that $\xi_{MAX} \le \frac{l}{2}$, then the equation of the body is

$$H = H_{MAX} C(t - t^n), \tag{5.10}$$

where

$$t \equiv 1 - \xi,$$

and H is the radius of the body, with n, H_{MAX}, and C as parameters.

If $\xi_{MAX} \ge \frac{l}{2}$, then

$$H = H_{MAX} C(\xi - \xi^n). \tag{5.11}$$

Since $A(x) = \pi H^2$ and $' \equiv \frac{d}{dx}$

$$A'' = 2\pi\left(H'^2 + HH''\right),$$

then

$$A'' = \frac{2\pi C^2 H_{MAX}^2}{l^2}\left[(1 - n\lambda)^2 - n(n-1)(\lambda - \lambda^2)\right], \tag{5.12}$$

where

$$\lambda = t^{n-1}.$$

The appropriate expression for $\xi_{A''=0}$ for the case covered by (5.9) is therefore

$$t_{A''=0} = \left\{\frac{b \pm \sqrt{b^2 - 4a}}{2a}\right\}^{\frac{1}{n-1}}, \tag{5.13}$$

where

$$a = 2n^2 - n, \quad b = n^2 + n\cdot$$

For the case covered by (5.11), t in (5.13) is replaced by ξ. These formulas were checked against values obtained from Fig. 4(e) in [11]. The positive branch is taken for the (3.3) case and the negative for the (5.11) case. Note that the branch other than that for the shock-body intersection corresponds to the body intersection of the upstream limb of the sonic line.

In the table on page 4 in [11], values of n are given for the various bodies tested. Figure 5.7 compares the shock location ξ_{SHOCK} obtained from our theoretical result

$$\xi_{SHOCK} = \xi_{A''=0} \tag{5.14}$$

with experiment.

The comparison in Fig. 5.7 is affected by sting-induced viscous phenomena occurring on the rearward part of the bodies tested which are shown in the figure. Despite the test Reynolds numbers being of the order of 40×10^6, strong diffusion of the shock appears at its foot, presumably due to shock-boundary layer interaction and shock-induced separation. Krupp in [12] and Malmuth in [13] have computed this case with a TSBT code. Krupp in Fig. 18 of [12] apparently labels his $M_\infty = .999$ case $M_\infty = 1$. The computations of [13] were for $M_\infty = .99$. Krupp used certain empirical scalings in the definition of K, r stretching, and the pressure coefficent C_P when he did these calculations. As far as the shock position is concerned, both computations produce results that are nearly identical. They are shown by the solid square in Fig. 5.7. This *inviscid* result agrees better with the theory then the experiments, in accord with expectations. Note also that the data in [11] were not corrected for wind-tunnel wall interference. Figure 5.8 shows isomachs discussed in [13] that were used to define the shock position ξ_{SHOCK} in Fig. 5.7.

5.4. Transonic Wind-Tunnel Wall Interference

Another example of TSBT is wind-tunnel wall interference. As in the previous applications, there is a need to rapidly estimate the effects of the walls on the model to extrapolate the measured aerodynamic characteristics to the free field (i.e., the free flight configuration of the walls absent). The same duality exists as discussed previously. Pure CFD models have been used. Since large parametric sets are tested, a need exists for an inexpensive complement to these simulations to assess interference quickly. CAN approaches are a promising candidate to fill this requirement.

References [13] and [14] describe CAN approaches to treat slender airplane configurations and high aspect ratio wings as well as procedures in which the asymptotics can be integrated with experimental measurements to access wall interference. For both the slender body and high aspect ratio cases, the wall interference is obtained by a systematic asymptotic expansion procedure. Each

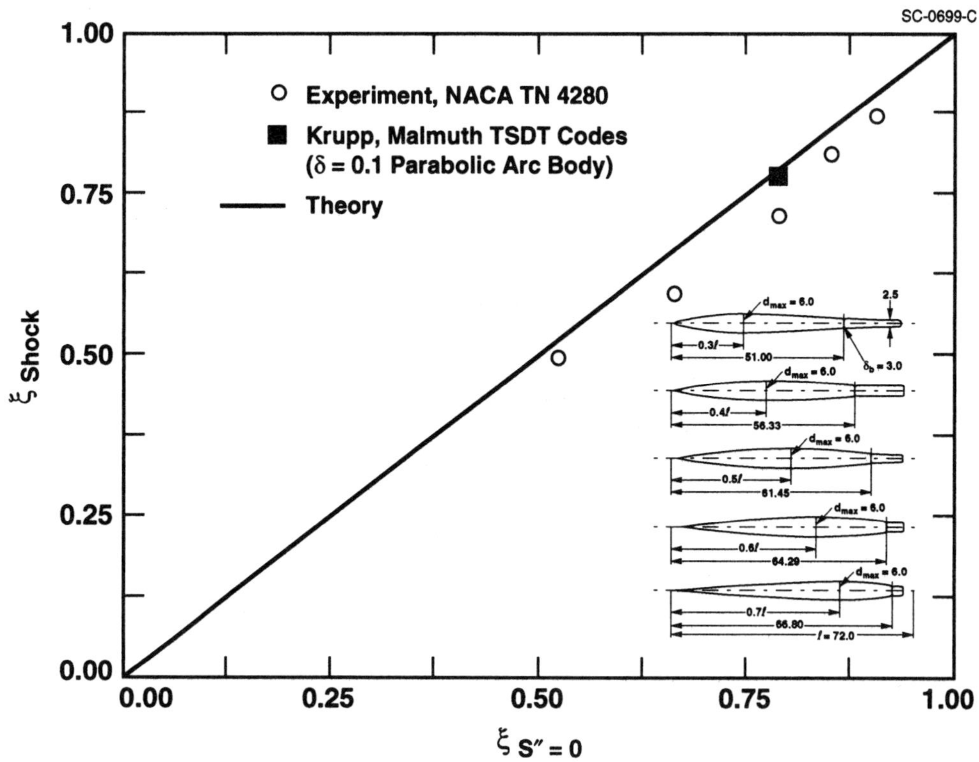

FIG. 5.7. *Comparison of shock locations from* (5.9) *with data of* [11].

is represented by a secondary approximation within a Karman–Guderley (KG) TSBT framework.

In the slender case, if the airplane model is represented as the surface

$$r = \delta F(x, \theta)\,, \tag{5.15}$$

within the coordinate system indicated in Fig. 5.9, with $\delta =$ the characteristic thickness ratio and overbars representing dimensional quantities, the asymptotic expansion of the velocity potential Φ in terms of the freestream speed U is

$$\frac{\Phi}{U} = x + \delta^2 \phi(x, \tilde{r}, \theta; K, H, A) + \cdots\,, \tag{5.16}$$

which holds for the KG outer limit,

$$x, \tilde{r} = \delta r, \theta, K = \left(1 - M_\infty^2\right)/\delta^2, H = h\delta/c, A = \alpha/\delta \text{ fixed as } \delta \to 0, \tag{5.17}$$

where $M_\infty =$ freestream Mach number, $K =$ KG similarity parameter, $H =$ scaled height of control surface, $A =$ incidence parameter. For (5.17), the pressure formula valid on the interface

$$C_P = -2\delta^2 \phi_x \tag{5.18}$$

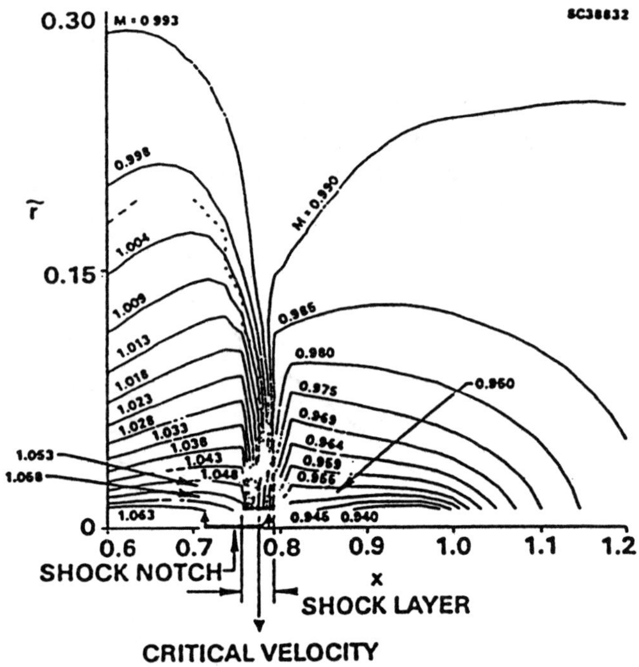

FIG. 5.8. *Isomachs for $M = .99$ flow over parabolic arc body, $\delta = .1$.*

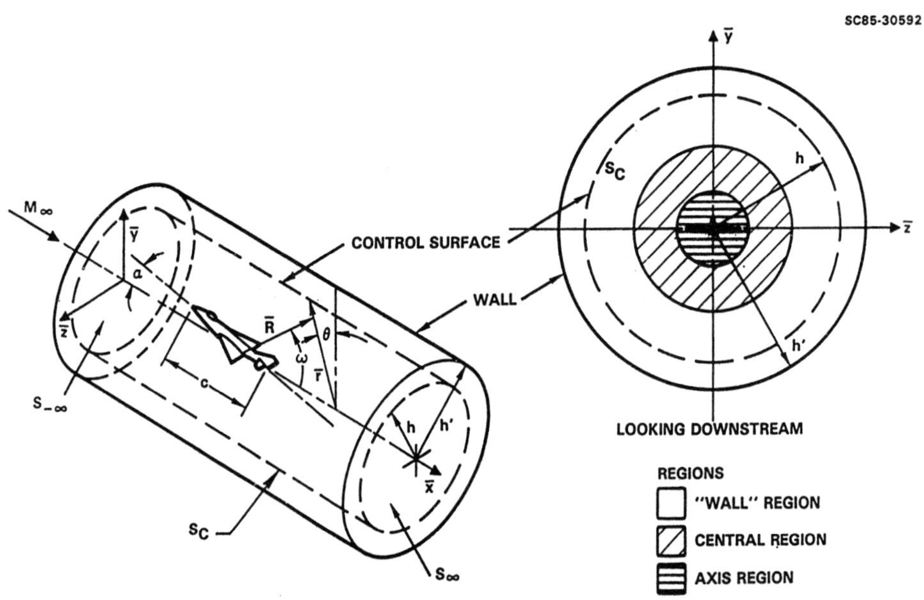

FIG. 5.9. *Confined slender airplane model.*

leads to a (primary) KG formulation for ϕ.

Secondary expansions for ϕ can be derived in the limit of $H \to \infty$. Three asymptotic decks can be identified for this limit. Shown schematically in Fig.

5.9 is an inner deck where cross flow gradients dominate, a central region that is weakly perturbed by the walls or interface deviations from free field conditions and a wall layer in which the model appears as a multipole in Prandtl–Glauert flow for "classical" Dirichlet and Neumann conditions associated with free jet and solid tunnel walls. Nonclassical conditions associated with slotted and porous walls can be handled by assuming that pressures are specified on a cylindrical "interface" surface near the walls. Matching procedures that are detailed in [13] and [15] provide the effect of reflections of the multipole in the interface for the classical case and give far field boundary conditions for a linearly perturbed KG problem for the interference. These are generalized to treat the nonclassical case as well in [13]. The boundary conditions as $\tilde{r} \to 0$ are obtained from matching with the inner deck or Axis layer. From these considerations, the boundary value problem **P** for the interference potential ϕ_1 representing the wall induced perturbations on the ϕ_0 free field flow in the central region is
P:

$$\text{(5.19a)} \quad [K_0^* - (\gamma+1)\phi_{0_x}]\,\phi_{1_{xx}} - (\gamma+1)\phi_{0_x}\phi_{1_x} + \frac{1}{\tilde{r}}(\tilde{r}\phi_{1_{\tilde{r}}})_{\tilde{r}} = -K_1^*\phi_{0_{xx}},$$

$$\text{(5.19b)} \quad \phi_{1_{\tilde{r}}}(x,0) = 0,$$

$$\text{(5.19c)} \quad \phi_1 \simeq \alpha_0 R^2 P_2(\cos\omega) + \alpha_1 R\cos\omega + \alpha_2 \quad \text{as } R \to \infty,$$

where $P_2(\cos\omega)$ is a Legendre polynomial, $R^2 = \frac{x^2}{K_0^*} + \tilde{r}^2$,

$$\text{(5.19d, e, f)} \quad \alpha_0 = \mathcal{C}_0 + \mathcal{D}_0, \quad \alpha_1 = B_0\alpha_0, \quad \alpha_2 = \alpha_0 C,$$

$$\text{(5.19g)} \quad 4\pi B_0 = -S(1) + \int_0^1 S(x)dx + \pi(\gamma+1)\int_{-\infty}^{\infty} dx \int_0^{\infty} r^2\phi_{0_x}^2 d\tilde{r},$$

$K_0^* = \frac{1-M_\infty^2}{\delta^2}$ = the free field similarity parameter and K_1^* = the perturbation of this parameter that can be adjusted for tunnel Mach number corrections. The constant C is the strength of the quadrupole part of the ϕ_0 far field, and $C_p(x,\theta;H) = f_x$, with $S(x) = \frac{A'(x)}{2\pi}$, where $A(x)$ is the streamwise area progression. The 0th order ϕ_0 free field "base" flow solves a KG problem for which $S(x)$ is the source strength of the equivalent body of revolution (EBR) on the x axis. The quantities $\mathcal{C}_0$ and $\mathcal{D}_0$ in (5.19d) are given by

$$\text{(5.19h)} \quad \mathcal{C}_0 = -\frac{1}{\pi}\int_0^{\infty}\frac{k^2dk}{I_0(k)}\int_0^{\infty} f(X)\cos kX\,dX,$$

$$\text{(5.19i)} \quad \mathcal{D}_0 = \frac{-S(1)}{2\pi^2\sqrt{K_0^*}}\int_0^{\infty}\frac{k^2K_0(k)}{I_0(k)}dk\,.$$

Here, ω is the scaled analogue of the polar angle defined in Fig. 5.9, i.e., $\omega = \cos^{-1} X/R$, with $X = x/\sqrt{K_0^*}$, and K_0 and I_0 are modified Bessel functions.

In the numerics, the third term in (5.19c) is neglected.

These equations are appropriate to pressure interface (f) data that have axial streamwise symmetry about $x = 0$. More general interface distributions are treated in [13]. For the free jet case, $\mathcal{C}_0 = 0$ in (5.19e). Solid wall conditions are modeled by making $\alpha_0 = b_0' = b_0 S(1)/\sqrt{K_0^*}$, and $\alpha_1 = 8\pi B_0 b_0$, with $b_0 = .063409$.

In addition to the boundary conditions given by (5.19b) and (5.19c), the Problem **P** is subject to the appropriate perturbation of the shock conditions. These are satisfied on the 0th order flow shocks. Since the interference flow ϕ_1 (1st order) boundary value problem is only sensitive to $A(x)$ (Area Rule for Interference), it can be used to rapidly assess interference for slender but realistic 3D configurations such as missiles and fighters, providing the angles of attack are moderate enough to warrant the application of this theory. References [16] and [17] provide other asymptotic frameworks to handle higher incidences.

Equation (5.19a) and its high aspect ratio analogue are "variational" equations of the interference flow. They are linear perturbations on the KG base flow with discontinuous coefficients associated with the shocks. Near the shock, they can be modeled by a generalized Lavrentef–Bitsadze equation. The variational equations are of mixed type and accordingly, modifications of the classical successive overrelaxation (SLOR) type sensitive differencing algorithms originally developed in [18] were used to treat them. The procedures used are generalizations employed by Small [19]. Because the coefficients of the second derivatives of ϕ_1 are frozen, the CPU time is an order of magnitude less than that for a KG solver. In contrast to the KG base flow, the shock is fitted rather than captured and the fact that the Rankine–Hugoniot jump perturbations can be linearized on the 0th order captured surface motivates the use of an inscribing "notch" on which the perturbation jumps serve as internal boundary conditions for the interference perturbation potential ϕ_0. Special methods used to couple the jumps along the notch with the SLOR method are described in [13]. For high aspect ratio shapes, an added technique is required, since there is an additional coupling of the shock notch-induced perturbations with the circulation of the interference flow, Γ_1. This motivated the application of convergence acceleration methods developed by Hafez and Cheng in [20]. Another feature of the numerical method involves subtracting off the far field singularity given by (5.19c). This allows homogeneous Dirichlet conditions to be imposed at the outer boundaries of the computational domain and regularizes the dependent variables by avoiding their scaling with the size of the domain. Convergence considerations associated with the size of the domain are thereby simplified. This regularization procedure introduces forcing terms in the variational equations.

For the slender body case, an important accuracy issue is the discretizations of the boundary conditions at $\tilde{r}$=0. Another feature of the numerical methods used in both our KG and variational equation solvers is the use of the Axis layer asymptotics in the boundary operators. We believe that this gives enhanced accuracy. This is an excellent example of the *asymptotic induced numerics* approach described in the introduction of this paper. A typical validation for an $H = O(1)$ simulation and our 0th order code is shown in Fig. 5.10. The data is for the B–3 body tested by Couch and Brooks at $M_\infty = .99$ [21]. In spite of the potent stagnation singularities for this blunt nose case, agreement with the data is excellent. Validations of this kind have provided encouraging evidence of the accuracy of the ϕ_0 solver.

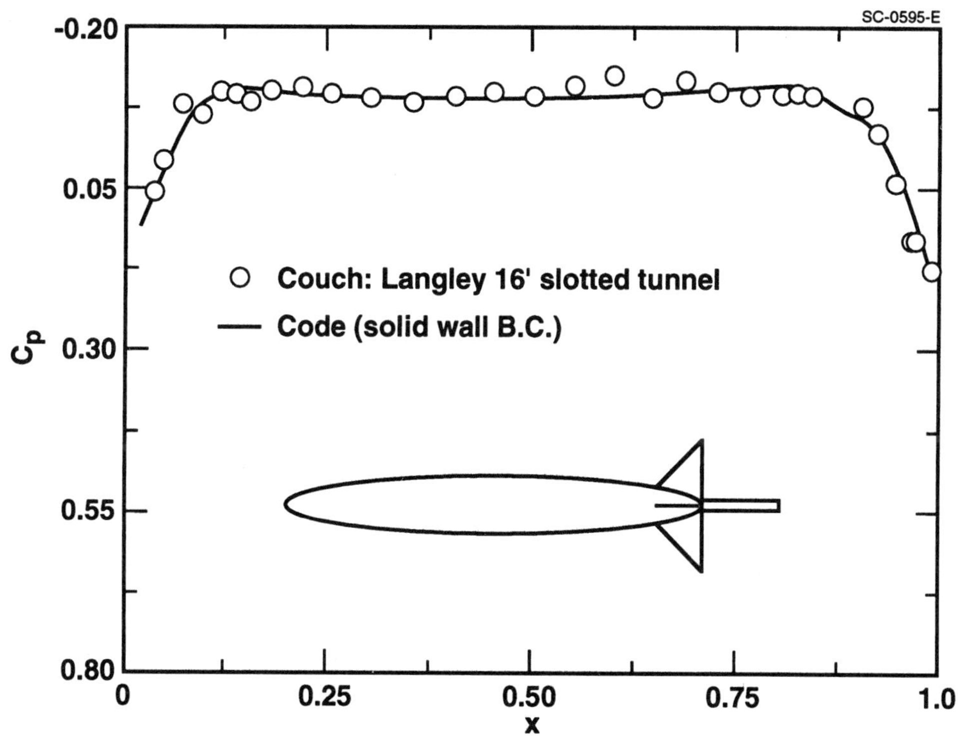

FIG. 5.10. *Comparison of theory of this paper with* [21].

In regard to the interference flow, Fig. 5.11 gives the Mach number dependence of the interference pressure distribution over a parabolic arc body confined by solid walls. For this part of the discussion, the parameter K signifies K_0^* introduced in (5.19a). Because the shock moves due to the change of the Mach number in the tunnel as compared to the free field, there is a spike in these distributions in the vicinity of the the free field shock location. Predictably, the intensity of this blip decreases while its width increases as the Mach number decreases. The qualitative variation of the interference pressure away from the location resembles that studied by Malmuth for incompressible flow in [22] by analytical methods, where the $H \to \infty$ theory was validated against an exact finite H solution. The interference mechanism is associated with a doublet reflection from the walls which gives a constant level in the large H limit, since there is an averaging effect due to the large distance of the wall from the model.

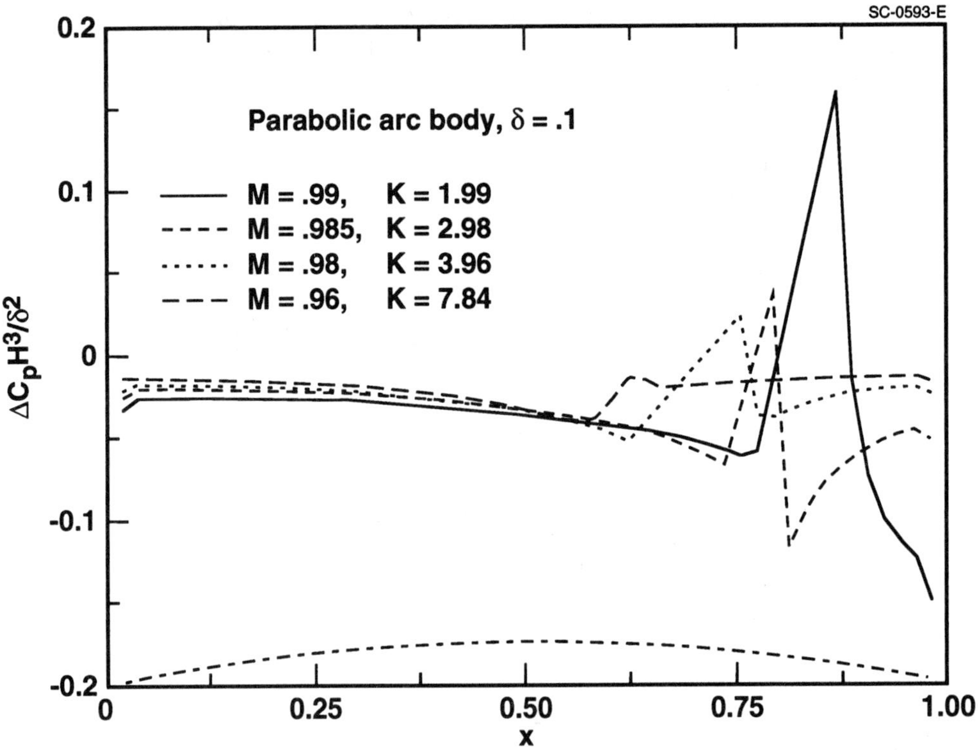

FIG. 5.11. *K dependence of reduced interference pressures for parabolic arc body.*

The associated effect on the the total surface pressures is shown in Fig. 5.12, which indicates the shock movement and the antisymmetry about the critical pressure level. This behavior is in accord with Prandtl's relation for normal shocks which was assumed to apply at the virtual intersection of the shock and the body. Natural outgrowths of the asymptotic representations that lead to the previously discussed boundary value problems are the similitudes of the interference pressure and drag. Matching of the three decks determines "gauge" functions in the expansions that appropriately order the terms. The ordinates and the abscissae of Fig. 5.11 and a plot of the interference drag given in Fig. 5.13 indicates this scaling which is consistent with that determined by Goethert in [23] using nonasymptotic procedures. An interference thrust that increases with Mach number is evident in Fig. 5.13. Although the walls provide an increase in Mach number and drag at the lower transonic Mach numbers due to their constrictive effect on the free field streamlines, this tendency is offset at the higher Mach numbers, presumably due to the post-shock compressions on the forward facing parts of the afterbody. These increase in intensity and

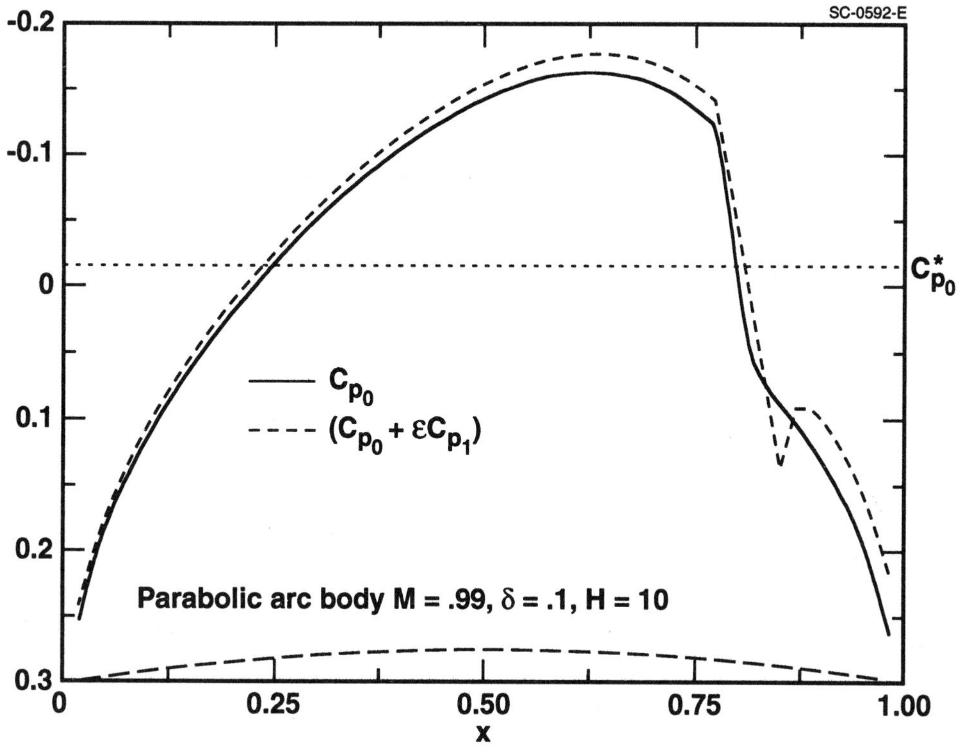

FIG. 5.12. *Comparison of* 0*th order and total* C_P, $H = 10+$, *parabolic body,* $M_\infty = .99$, $\delta = .1$, $K = 1.99$.

move rearward with increasing M_∞.

5.5. Web Levitation and Foil Bearings

A common occurrence in industrial processes and lubrication technology is the transport of single and multiple continuous sheets (called hereinafter webs) through rollers and over other guide, folding, forming, and cutting elements. The applications are diverse, including film and tape manufacture as well as paper production and tape transport devices used in the computing and entertainment industries (associated with magnetic recording and tape drives). Commercial and defense aspects relate to yet another application involving foil bearings in which the web has flexural rigidity. The configuration of the foil moving over the solid surface is directly analogous to the aforementioned web transport applications in which the web usually has unlimited flexibility. A typical foil bearing configuration is shown in Fig. 5.14.

To augment lubrication, foil bearings frequently use *external pressurization* to inject fluid in the region of potential contact between sliding surfaces. Slots, holes, and grooves on the cylinder C shown in Fig. 5.14 can be used to convect the lubricating fluid in the region to be lubricated. A typical example of gas as the lubricating fluid is the gas bearing. Although there is a rich literature

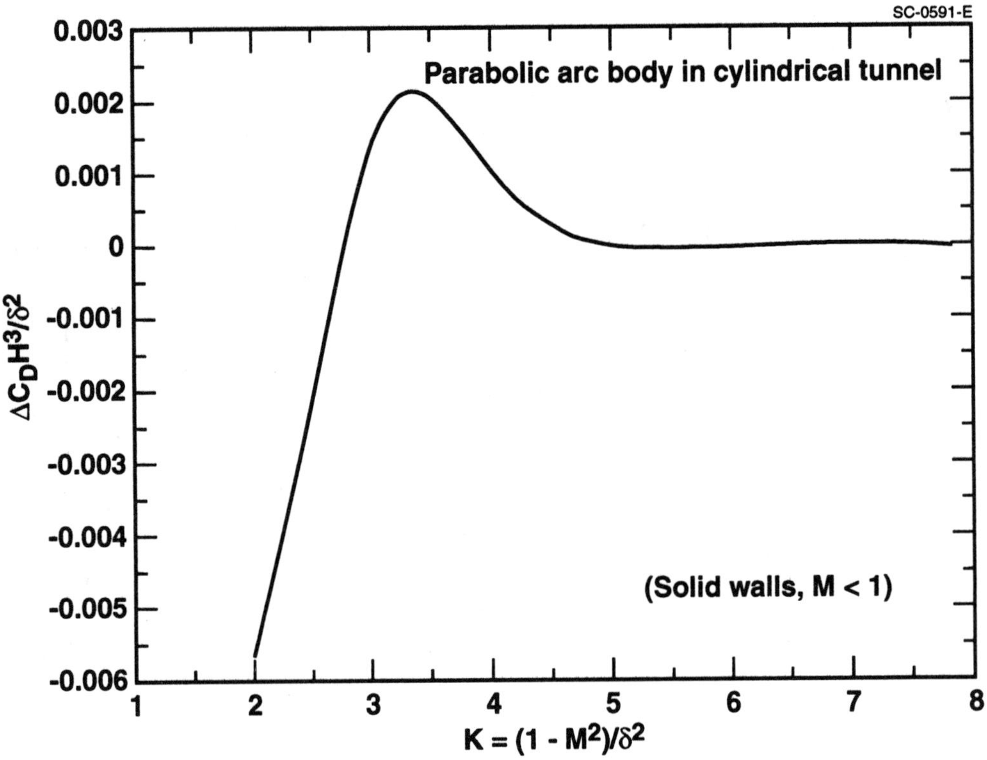

FIG. 5.13. *Normalized interference drag* $\frac{\Delta C_D H^3}{\delta^3}$ *as a function of the transonic similarity parameter* K.

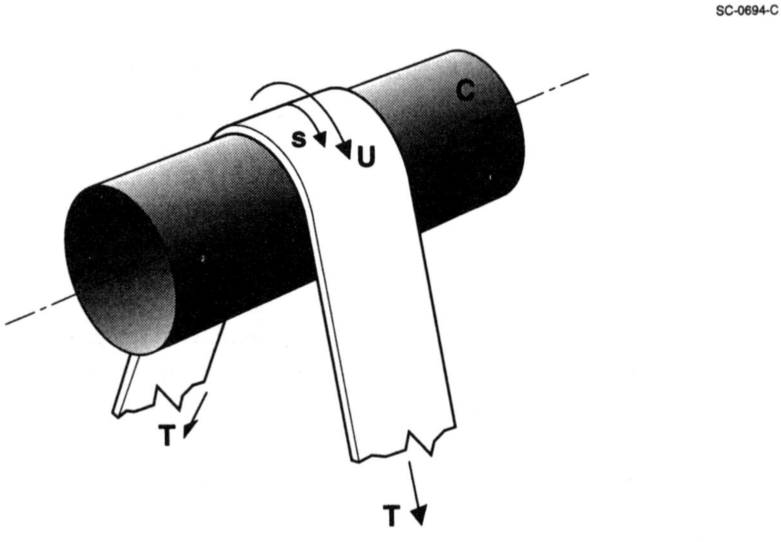

FIG. 5.14. *Schematic of foil bearing.*

[24]–[33] on these applications, theoretical treatment of the underlying fluid mechanics seems to be limited to lubrication theory. Eschel and Elrod [26] treated foil bearings without external pressurization. Singular perturbations and matching can be applied for the externally pressurized case. Our approach for this is given in which large tension approximations are applied for the first time to treat arbitrary variations of pressure, and velocity on the slotted cylinder bearing surface using this consistent approximation scheme.

Figure 5.14 shows the overall physical flow system to be considered. The web in this figure will be assumed to be traveling at the velocity U perpendicular to the symmetry axis of the bearing cylinder C. Figure 5.15 shows a cross-section of the web flow over the cylinder C in which external pressurization is applied to the gap between the web and C from the plenum P through the slot system shown. A blow-up of the gap region and an overall view of the major flows involved in this application are shown in Fig. 5.16. It is important to understand the fluid dynamical processes associated with gas-dynamic lubrication by levitation in this kind of foil bearing. Clearly, diverse length scales interact. Regions I–IV shown in Fig. 5.16 are coupled. Here Region I is the thin gap lubrication zone; Region II is the outer region in which circulations are induced by the moving web. Other circulations and eddies can be produced in Region IV and there is a strong interaction between Region I and the plenum (Region III).

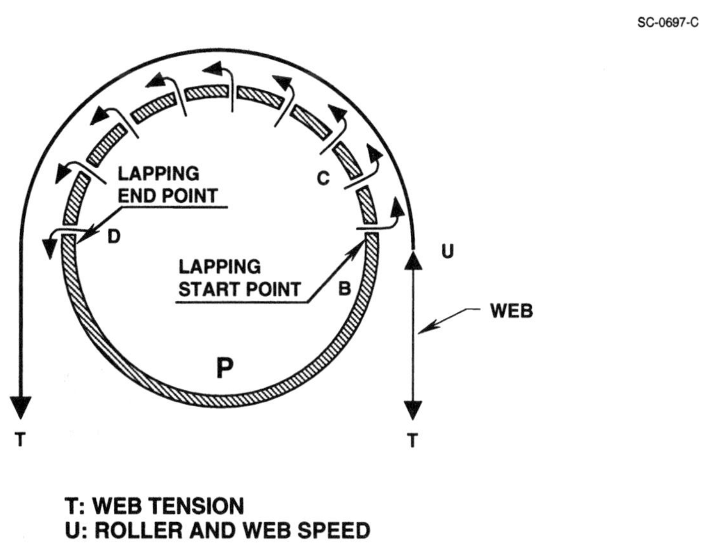

FIG. 5.15. *Schematic of airbar.*

As previously indicated, the global flow structure in the narrow gap Region I in Fig. 5.16 can be approximated by lubrication theory. If $\vec{q} = \overline{q}_r \vec{1}_r + \overline{q}_\theta \vec{1}_\theta$, asymptotic expansions can be used to define a consistent approximation scheme to solve the elastohydrodynamic interaction of the web with the flow field. If the Re based on an $O(1)$ length $= \epsilon^{-1}$, then if $\overline{p}$ is the dimensional

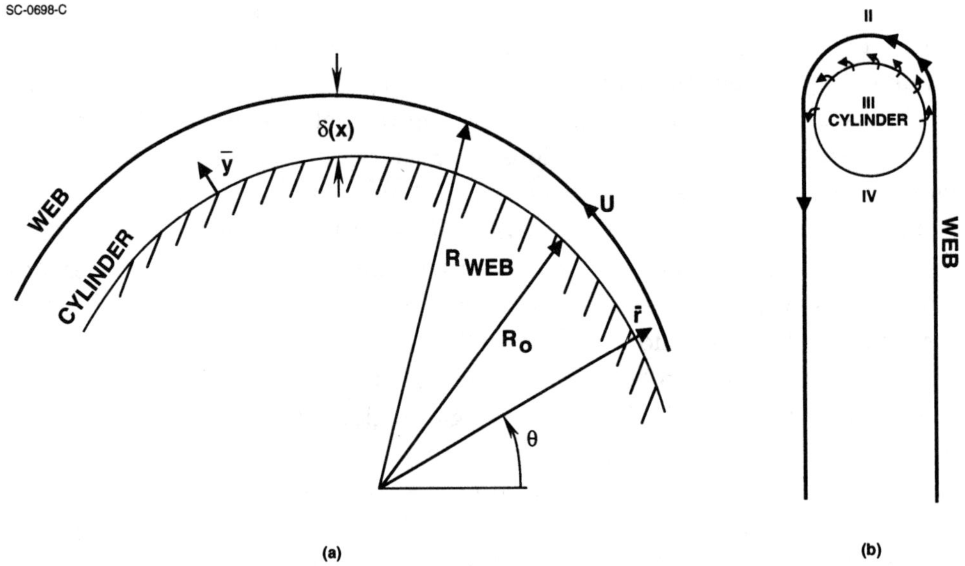

FIG. 5.16. *Narrow gap flow Region I in relation to others.*

pressure, $\rho =$ density, and $p = \frac{\overline{p}}{\rho U^2}$, the expansions

(5.20a) $$q_\theta(x, \tilde{y}; \epsilon) = u_0(x, \tilde{y}) + \cdots,$$

(5.20b) $$q_r(x, \tilde{y}; \epsilon) = \epsilon v_0(x, \tilde{y}) + \cdots,$$

(5.20c) $$p(x, \tilde{y}; \epsilon) = \frac{p_0(x, \tilde{y})}{\epsilon} + \cdots,$$

for the distinguished limit

(5.20d) $$x, \tilde{y} = y/\epsilon \text{ fixed as } \epsilon = \frac{1}{Re} \to 0,$$

using the coordinate system shown in Fig. 5.16, when substituted into the incompressible Navier–Stokes equations lead to the lubrication theory equations. Here, the stretched coordinate $\tilde{y}$ is used to keep the gap region on view in the limit (5.20d).

Boundary conditions for these equations are related to appropriate matching with the wall boundary conditions on the cylinder surface and the local slot or hole flow fields as well as dynamical surface traction boundary conditions on the web.

A model has been constructed allowing for arbitrary distributions of normal and tangential velocity components as well as pressure on the ventilated surface. Details of the analysis are given in [34]. Actually, the connection

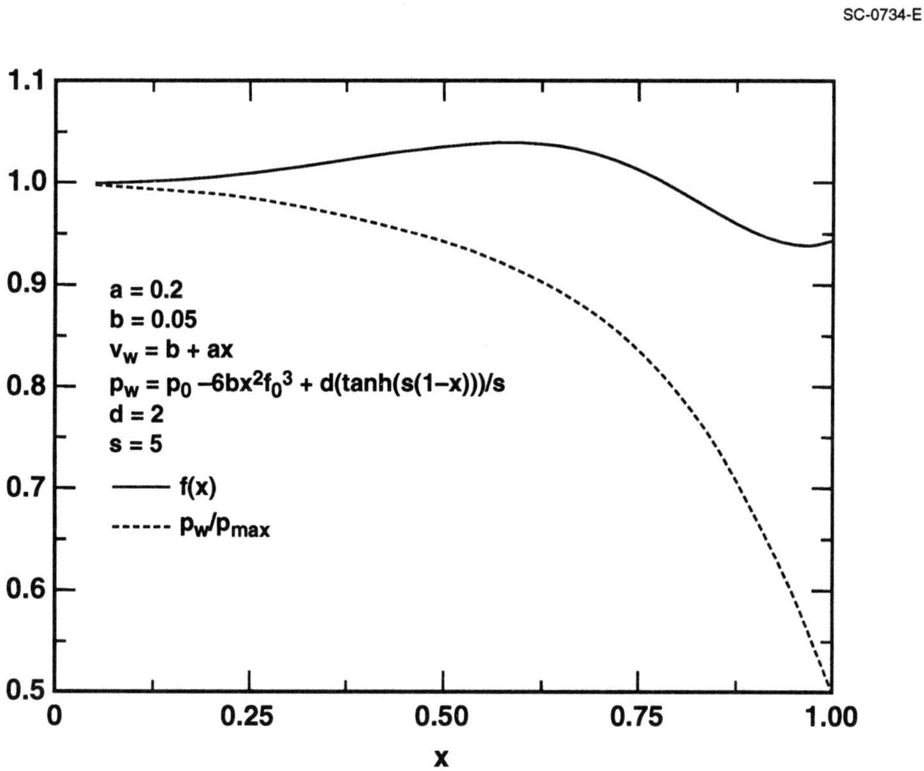

FIG. 5.17. *Web height variation,* $x = \frac{\theta}{\theta_{MAX}}$, $f_0(0) = 1$, $u_{0_W} = 0$, $f'(0) \neq 0$.

between the slot density and geometry with these distributions needs to be established. This connection is important in determining power requirements and the economics discussed earlier.

Presumably, asymptotic matching can be used to connect the hole or slot geometry and the "inner" slot flow with the outer thin gap Region I lubrication flow. Region I, in turn, could provide outer boundary conditions for the local slot flow problem.

The solution of the Region I flow problem discussed in [34] is

$$\text{(5.21a)}\quad u_0 = u_{0_w}(x) + p'_{0_w}(x)\frac{\tilde{y}^2}{2} + \left\{\frac{1 - u_{0_w}(x)}{f(x)} - \frac{p'_{0_w}(x)f(x)}{2}\right\}\tilde{y},$$

$$v_0 = v_{0_w}(x) - u'_{0_w}(x)\tilde{y} - p''_{0_w}\frac{\tilde{y}^3}{6}$$

$$\text{(5.21b)}\quad + \left\{\frac{u'_{0_w}(x)}{f(x)} + \frac{p''_{0_w}(x)f(x)}{2} + \left[\frac{1 - u_{0_w}(x)}{f^2(x)} + \frac{p'_{0_w}(x)f'(x)}{2}\right]\right\}\frac{\tilde{y}^2}{2},$$

$$\text{(5.21c)}\quad p_0 = p_{0_w}(x),$$

where subscript w quantities refer to the arbitrary ventilated cylinder surface distributions mentioned previously. Denoting the wall deflection $\delta(x) = \epsilon f(x)$ in the coordinate system shown in Fig. 5.16, use of the surface traction and no-slip boundary condition leads to a first order nonlinear differential equation for $f(x)$ which can be integrated to give the cubic

$$Af + Bf^3 - C = 0, \tag{5.22}$$

where A, B, and C are functions of the wall distribution functions and a scale height f_0 determine by matching with end layers near the the web entry and exit wrap-around points and an integral conservation law. A typical solution of the cubic (5.22) is plotted in Fig. 5.17 which shows the interplay between the blowing interface pressures and uniformity of levitation. Variations such as these have demonstrated the profound effect of the ambient decay of pressure on levitation uniformity which interacts with lubrication effectiveness.

5.6. Conclusions

Various examples have been discussed that illustrate the power of CAN for nonlinear problems. Although the transonic equivalence and area rule have been known for many years the combination of them with modern computational techniques has been untapped until recently. This combination is quite useful for preliminary design and understanding the basic physics of the flow. Future extensions of transonic slender body theory CAN can test the applicablity of the shock location rule for bodies of revolution to asymmetric bodies. Examples relating to wind-tunnel wall interference and foil bearings provide useful insights and generalizations.

References

[1] J. D. Cole, *Perturbation Methods in Applied Mathematics*, Blaisdell Publishing Company, Waltham, MA, 1968.

[2] P.A. Lagerstrom, *Matched Asymptotic Expansions, Ideas and Techniques*, Springer-Verlag, New York, 1988.

[3] M. D. Van Dyke, *Perturbation Methods in Fluid Mechanics*, Parabolic Press, Stanford, CT, 1975.

[4] J. Cole and J. Kevorkian, *Perturbation Methods in Applied Mathematics*, Springer-Verlag, New York, 1980.

[5] K. Oswatitsch and F. Keune, *Ein Aquivalensatz fur Nichtfangstelle Fluge Keiner Spannwelte in Schallnaher Strömung*, Z. Flugwissenshaften, 3(2) (1955), s. 29–46.

[6] J. Cole and P. Cook, *Transonic Aerodynamics*, North-Holland, Amsterdam, 1986.

[7] R. T. Whitcomb, *A Study of Zero Lift Design Rise Characteristics of Wing-Body Combinations Near the Speed of Sound*, NACA Rpt. 1273, 1952.

[8] R. T. Jones, *Theory of Wing-Body Drag at Supersonic Speeds*, NACA Rpt. 1318, September 1956.

[9] K. Rajagopal, N. Malmuth, and W. Lick, *Application of Transonic Slender Body Theory to Various Shapes*, AIAA Paper 88–0005, presented at the 26th Aerospace Sciences Meeting, Reno, Nevada, January 11–14, 1988. AIAA J., 27(8) (1989), pp. 1220–1229.

[10] J. D. Cole and N. Malmuth, *Shock wave location on a slender transonic body of revolution*, Mech. Res. Comm., 10(6) (1989), pp. 353–335.

[11] J. McDevitt and R. Taylor, *Pressure Distributions at Transonic Speeds for Slender Bodies Having Various Axial Locations of of Maximum Diameter*, NACA Technical Note TN 4280, July 1958.

[12] J. A. Krupp and E. M. Murman, *Computation of transonic flows past lifting airfoils and slender bodies*, AIAA J., 10(7) (1972), pp. 880–886.

[13] N. D. Malmuth, C. C. Wu, H. Jafroudi, R. Mclachlan, J. D. Cole, and R. Sahu, *Asymptotic Theory of Wind Tunnel Wall Interference*, AEDC Final Report for Contract F40600-84-C-0010, period March 30, 1984 through July 30, 1990, AEDC-TR-91-24, December 1991.

[14] N. D. Malmuth, H. Jafroudi, C. Wu, R. Mclachlan, and J. Cole, *Asymptotic Methods Applied to Transonic Wall Interference*, AIAA Paper 91–1712, presented at the AIAA 22nd Fluid and Plasma Dynamics Conference, June 22–26, 1991, Honolulu, Hawaii. AIAA J., 1993, in press.

[15] N. D. Malmuth and J. D. Cole, *Study of Asymptotic Theory of Transonic Wind Tunnel Interference*, Final Report for Period May 30, 1982, through August 30, 1983, Contract No. F40600–82–C–0005, Arnold Engineering Development Center/DOS Report AEDC–TR–84–8, Tullahoma, Tennessee, May 1984.

[16] J. D. Cole and N. D. Malmuth, *Wave Drag Due to Lift of Transonic Airplanes*, Transonic Symposium: Theory, Application and Experiment, NASA CP–3020, 1988, pp. 293–308.

[17] J. D. Cole and L. P. Cook, *Some Asymptotic Problems of Transonic Flow Theory*, presented at the Symposium Transsonicum III, Göttingen, West Germany, May 1988.

[18] E. M. Murman and J. D. Cole, *Calculation of plane transonic flows*, AIAA J., 9 (1971), pp. 114–121.

[19] R. D. Small, *Studies in Transonic Flow* VI, *Calculation of a Transonic Lifting Line Theory*, UCLA Report UCLA–ENG–7836, April 1978.

[20] M. M. Hafez and H. K. Cheng, *Convergence Acceleration of Relaxation Solutions for Transonic Computations*, AIAA Paper 75–71, 1975.

[21] L. M. Couch and C. W. Brooks, Jr., *Effect of Blockage Ratio on Drag and Pressure Distributions for Bodies of Revolution at Transonic Speeds*, NASA TN D-7331, 1973.

[22] N. D. Malmuth, *An asymptotic theory of wind tunnel wall interference on subsonic slender bodies*, J. Fluid Mech., 177 (1987), pp. 19–35.

[23] B. H. Goethert, *Transonic Wind Tunnel Testing*, Pergamon, New York, 1961.

[24] D. Fuller,*Theory and Practice of Lubrication for Engineers*, John Wiley, New York, 1956.

[25] H. Block and J. J. vanRossum, *The foil bearing—A new departure in hydrodynamic lubrication*, Lubrication Engineering, December 1953, pp. 316–320.

[26] A. Eshel and H. G. Elrod, Jr., *The theory of the infinitely wide, perfectly flexible, self-acting foil bearing*, Trans. ASME, Journal of Lubrication Technology, December 1965, pp. 831–836.

[27] L. Licht, *An experimental study of elastohydrodynamic lubrication of foil bearings*, Trans. ASME, Journal of Lubrication Technology, July 1968, pp. 199–220.

[28] E. J. Barlow, *Derivation of governing equations for self-acting foil bearings*, Trans. ASME, Journal of Lubrication Technology, January 1968, pp. 334–340.

[29] T. B. Barnum and H. G. Elrod, Jr., *A theoretical study of the dynamic behavior of foil bearings*, Trans. ASME, Journal of Lubrication Technology, January 1971, pp. 133–142.

[30] M. Wildmann, *Foil bearings*, Trans. ASME, Journal of Lubrication Technology, January 1969, pp. 37–44.

[31] W. Gross et al., *Fluid Film Lubrication*, John Wiley, New York, 1980.

[32] S. M. Vogel and J. L. Groom, *White light interferometry of elastohydrodynamic lubrication of foil bearings*, IBM J. Res. Dev., November 1974, pp. 521–528.

[33] C. A. Lacey and F. E. Talke, *A tightly coupled numerical foil bearing solution*, IEEE Trans. Magnetics, 26(6) (1990), pp. 3039–3043.

[34] N. Malmuth, D. Hobson, and L. Bivins, *Aerodynamics of Web Handling*, Rockwell Science Center Internal Report SC–071.90, September 19, 1990.

Index